Computer Communications and Networks

For further volumes:
www.springer.com/series/4198

The Computer Communications and Networks series is a range of textbooks, monographs and handbooks. It sets out to provide students, researchers and non-specialists alike with a sure grounding in current knowledge, together with comprehensible access to the latest developments in computer communications and networking.

Emphasis is placed on clear and explanatory styles that support a tutorial approach, so that even the most complex of topics is presented in a lucid and intelligible manner.

K. Erciyes

Distributed Graph Algorithms for Computer Networks

 Springer

K. Erciyes
Computer Engineering Department
Izmir University
Uckuyular, Izmir, Turkey

Series Editor
A.J. Sammes
Centre for Forensic Computing
Cranfield University
Shrivenham campus
Swindon, UK

ISSN 1617-7975   Computer Communications and Networks
ISBN 978-1-4471-5850-9   ISBN 978-1-4471-5173-9 (eBook)
DOI 10.1007/978-1-4471-5173-9
Springer London Heidelberg New York Dordrecht

To the memories of Necdet Doğanata and Selçuk Erciyes, and all who believe in education

Preface

Distributed systems consisting of a number of autonomous computing elements connected over a communication network that cooperate to achieve common goals have shown an unprecedented growth in the last few decades, especially in the form of the Grid, the Cloud, mobile ad hoc networks, and wireless sensor networks. Design of algorithms for these systems, namely the distributed algorithms, has become an important research area of computer science, engineering, applied mathematics, and other disciplines as they pose different and usually more difficult problems than the sequential algorithms. A graph can be used to conveniently model a distributed system, and distributed graph algorithms or graph-theoretical distributed algorithms, in the context of this book, are considered as distributed algorithms that make use of some property of the graph that models the distributed system to solve a problem in such systems.

This book is about distributed graph algorithms as applied to computer networks with focus on implementation and hopefully without much sacrifice on the theory. It grew out of the need I have witnessed while teaching distributed systems and algorithms courses in the last two decades or so. The main observation was that although there were many books on distributed algorithms, graph theory, and ad hoc networks separately, there did not seem to be any book with detailed focus on the intersection of these three major areas of research. The second observation was the difficulty the students faced when implementing distributed algorithm code although the concepts and the idea of an algorithm in an abstract manner were perceived relatively more comfortably. For example, when and how to synchronize algorithms running on different computing nodes was one of the main difficulties. In this sense, we have attempted to provide algorithms in ready-to-be-coded format in most cases, showing minor details explicitly to aid the distributed algorithm designer and implementor.

The book is divided into three parts. After reviewing the background, Part I provides a review of the fundamental and better known distributed graph algorithms. Part II describes the core concepts of distributed graph algorithms that have wide range of applications in computer networks in an abstract manner, without considering the application environment. However, in Part III, we focus ourselves on ad hoc wireless networks and show how some of the algorithms we have investigated can be modified for this environment.

The layout of each chapter is kept quite uniform for ease of reading. Each chapter starts with an introduction describing the problem shortly by showing its possible applications in computer networks. The problem is then stated formally, and examples are provided in most of the cases. We then provide a list of algorithms usually starting by a sequential one to aid understanding the problem better. The distributed algorithms shown may be well established if they exist and sometimes algorithms that have been recently published as articles are described with examples if they have profound effect on the solution of the problem.

An algorithm is first introduced conceptually, and then, its pseudocode is given and described in detail. We provide similar simple graph templates to show the steps of the implementation of the algorithm and then provide analysis of its time and message complexity. Proof of correctness is given only when this does not seem obvious or, on the contrary, a reference is given for the proof if this requires lengthy analysis. The chapter concludes by the Chapter Notes section, which usually emphasizes main points, compares the described algorithms, and also provides a contemporary bibliographic review of the topic with open research areas where applicable. This style is repeated throughout the book for all chapters. Exercises at the end of chapters are usually in the form of small programming projects in line with the main goal of the book, which is to describe how to implement distributed algorithms.

There are few aspects of the book worth mentioning. Firstly, many self-stabilizing algorithms are included, some being very recent, for most of the topics covered in Part II. There are few algorithms, again in Part II, that are new and have not been published elsewhere. Also, an updated survey of the topic covered is provided for all chapters. Finally, a simple simulator we have designed, implemented, and used while teaching distributed algorithm courses is included as the final chapter, and its source code is given in Appendix B.

The intended audience for this book are the graduate students and researchers of computer science and mathematics and engineering or any person with basic background in discrete mathematics, algorithms, and computer networks.

I would like to thank graduate students at Ege University, University of California Davis, California State University San Marcos and senior students at Izmir University who have taken the distributed algorithms courses, sometimes under slightly different names, for their valuable feedback when parts of the material covered in the book was presented during lectures. I would like to thank Aysegul Alaybeyoglu, Deniz Cokuslu, Orhan Dagdeviren, and Jukka Suomela for their review of some chapters and valuable comments. I would also like to thank Springer editors Wayne Wheeler and Simon Rees for their continuous support during the course of this project and Donatas Akmanavičius for the final editing process.

Izmir, Turkey   K. Erciyes

Contents

Acronyms

AoA	Angle of Arrival
APSP	All Pairs Shortest Paths
ASSIST	A Simple Simulator based on Threads
BFS	Breadth First Search
CDS	Connected Dominating Set
DFS	Depth First Search
DS	Dominating Set
DT	Delaunay Triangulation
EKF	Extended Kalman Filter
FSM	Finite State Machine
GG	Gabriel Graph
IS	Independent Set
KF	Kalman Filter
k-NNG	k-Nearest Neighbor Graph
MaxIS	Maximum Independent Set
MaxM	Maximum Matching
MaxWM	Maximum Weighted Matching
MCDS	Minimal Connected Dominating Set
MCVC	Minimal Connected Vertex Cover
MCWVC	Minimal Connected Weighted Vertex Cover
MDS	Minimal Dominating Set
MinCDS	Minimum Connected Dominating Set
MinCVC	Minimum Connected Vertex Cover
MinCWVC	Minimum Connected Weighted Vertex Cover
MinDS	Minimum Dominating Set
MinVC	Minimum Vertex Cover
MinWVC	Minimum Weighted Vertex Cover
MIS	Maximal Weighted Matching
MM	Maximal Matching
MST	Minimum Spanning Tree
MVC	Minimal Vertex Cover

MWM	Maximal Weighted Matching
MWOE	Minimum Weight Outgoing Edge
MWVC	Minimal Weighted Vertex Cover
NNG	Nearest-Neighbor Graph
PF	Particle Filter
QUDG	Quasi Unit Disk Graph
RNG	Relative Neighborhood Graph
RSSI	Received Signal Strength Indicator
SSSP	Single-Source Shortest Paths
TDoA	Time Difference of Arrival
UDG	Unit Disk Graph
VC	Vertex Cover
YG	Yao Graph

Chapter 1
Introduction

Abstract A distributed system consists of a set of computational nodes connected by a communication network that cooperate to accomplish common tasks. In this chapter, we will review the benefits of using a distributed system, the architecture of a distributed system, and the challenges facing the designers.

1.1 Distributed Systems

The basic requirements from a distributed system are that the nodes should be autonomous so that they can work independently; the network should be connected, that is, any node should have a communication link directly or indirectly to any other node; and there should be a coordination mechanism for the nodes to cooperate to achieve common goals.

There are a number of benefits to be gained by utilizing distributed systems. One of the obvious advantages of using a distributed system is *resource sharing*. Access to a central resource has two disadvantages as this central site becomes a bottleneck for communications and also is a single point of failure. Distributing the resources such as the database and peripherals over a network overcomes these problems.

Resources and computation can be replicated at various sites providing fault tolerance as a replica may be substituted in the case of the dysfunctioning of a node. This type of fault tolerance is an important reason to employ distributed systems. It is also possible for the application to be inherently distributed such as bank transaction systems and airline reservation systems where employment of distributed systems is inevitable.

A distributed system can be modeled as a graph $G(V, E)$ conveniently where V is the set of vertices and E is the set of edges of G. The computing nodes of the distributed system are represented by the vertices of the graph, and an edge exists between the nodes if there is a communication link between them. Figure 1.1 displays a graph that represents a distributed system consisting of nodes numbered $1, \ldots, 10$. The first thing that may be noticed is that the graph is connected, providing a communication path between any pair of nodes. Many nodes are not directly connected to each other; therefore, they have to rely on their neighbor nodes to communicate with the other nodes of the network.

K. Erciyes, *Distributed Graph Algorithms for Computer Networks*,
Computer Communications and Networks, DOI 10.1007/978-1-4471-5173-9_1,
© Springer-Verlag London 2013

Fig. 1.1 A graph
representing a distributed
system

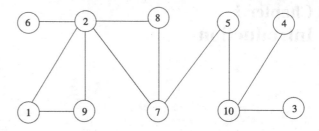

We will use graphs to represent distributed systems and show the execution of
a distributed algorithm in these graphs frequently. In this chapter, we will first describe platforms and models for distributed computing in Sects. 1.2 and 1.3 and
then describe the software architecture of a distributed system in Sect. 1.4. The
challenges in the design of distributed algorithms are reviewed in Sect. 1.5, and distributed graph algorithms are described in Sect. 1.6. Finally, we conclude by the
organization of the book.

1.2 Distributed Computing Platforms

Due to the recent technological advancements, in the last few decades, we have
witnessed diverse distributed system platforms such as the Grid, The Cloud, mobile
ad hoc networks, and wireless sensor networks that are described below.

1.2.1 The Grid

The Grid consists of loosely coupled, heterogeneous, and geographically dispersed
computing elements that are connected by a network acting together to perform
large tasks [3]. These computationally intensive scientific tasks may include various
applications such as seismic analysis, drug discovery, and bioinformatics problems.
Grid computing provides effective usage of the unused processing power and results
in decreased completion time for a task due to parallelization.

The size of a grid varies from a small network of workstations in a corpora-
tion to thousands of nodes across many networks and nations. Grids require general
software libraries called the *middleware* to accomplish coordination among a large
number of nodes that comprise them. *Resource discovery* is the process of finding
the location of the required resources such as the database tables in the Grid [2].
Resource allocation process, on the other hand, tries to map these resources to the
application requirements for the best performance. Both resource discovery and resource allocation are active research areas for the grids. An important problem with
the grids is that nodes may abort due to faults that may be difficult to find and take
necessary action due to the lack of central control. For this reason, fault tolerance

and also load balancing is another important research area in the grids [8]. Lack of central control and the need to provide access to a large number of users requires protection due to possible risks. The European Grid Infrastructure (EGI) is a grid for high-energy physics, earth observation, and biology applications [6], and in the United States, the National Grid (USNG) [9] is prototyping a computational grid for infrastructure and an access grid for people.

1.2.2 Cloud Computing

The *cloud computing* evolved from grid computing with the aim to deliver the computing as a service to the users by extending the object-oriented programming paradigm. Cloud computing provides computation, software applications, data access, data management, and storage for resources without requiring cloud users to know the location and other details of the computing infrastructure [7]. Grid computing may be included in the cloud or not depending on the type of application and users. Cloud computing and grid computing aim at scalability, and both use load balancing to accomplish scalability. In grid computing, a single task is divided into smaller tasks that are run on a number of processors to effectively use the available computing power, whereas in cloud computing, service offered to users is not restricted to processing power and includes website hosting, database support, etc. Cloud computing, in general, offers more services than the Grid.

1.2.3 Mobile Ad hoc Networks

A *wireless ad hoc network* is a decentralized network consisting of wireless nodes that do not rely on a predefined infrastructure such as routers or access points. Instead, each node participates in routing by forwarding data to other nodes regarding dynamically changing network topology. A *mobile ad hoc network* (*MANET*) is a network without any fixed structure formed for a purpose by mobile devices connected by wireless communication links. Each node of a MANET moves independently, forming a dynamic network that changes its topology continuously. Nodes of a MANET must be able to route any messages not destined to them; therefore, each node functions as a router. Examples of MANETs are the disaster relief operations, military networks, and vehicular ad hoc networks.

1.2.4 Wireless Sensor Networks

A *wireless sensor network* (*WSN*) consists of many small nodes of computing elements, each equipped with sensing and wireless communication capabilities. These networks can obtain data about their environment and transfer this data to a central node using *multi-hop communication* to be analyzed further. The WSNs have large application spectrum such as habitat monitoring, military surveillance, and target tracking [1]. WSNs form a large-scale distributed system and require scalable distributed algorithms to solve problems such as data aggregation, topology control, and routing.

1.3 Models

The basic models of a distributed system are the *message passing* and *shared-memory* models. In the message passing model, nodes of the distributed system communicate by messages only. Messages are communicated in rounds in synchronous message passing, where messages sent in round k are delivered to all recipients before messages in round $k + 1$ can be transferred. In asynchronous message passing, however, messages are assumed to eventually reach the destinations after unknown delays. Analyzing asynchronous message passing algorithms is more difficult than synchronous ones due to the uncertainties involved.

In shared-memory models, processes communicate by reading and writing to shared memory. Synchronization is an important issue also in shared-memory systems. *Distributed shared-memory* systems implement shared memory model over the message passing model to use the available shared memory software modules conveniently. Our analysis in this book is confined to message-passing distributed systems without any shared memory in general, except for some self-stabilizing algorithms, where it will be assumed that a process can read the values of the registers of its neighbors.

1.4 Software Architecture

The software modules at a node of a distributed computing system consist of the distributed algorithm that is the application software: the local operating system, the middleware, and the protocol stack as shown in Fig. 1.2. The operating system at each node is mainly responsible for resource management tasks such as file and memory management and local synchronization among local tasks. A distributed operating system, on the other hand, aims to provide global resource management, synchronization, and services to the users so that the users are not aware of the location of the service.

Instead of designing and implementing a distributed operating system from scratch, its tasks are usually handled by special software modules called the *middleware* targeting at the specific task at hand. The middleware layer is between the local operating system and the application software, and a software module in this layer performs a specific function that may be required by a number of applications. For example, a *synchronizer* is a middleware module that provides synchronization among application level processes, and any application that needs synchronization may use this module by invoking its interface routines.

Fig. 1.2 Software modules
of a network node

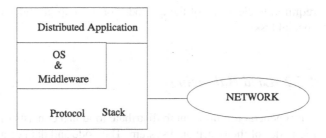

The protocol stack is responsible for the correct and timely delivery of messages between the nodes of the distributed system. Distributed systems do not have a common clock and therefore require synchronization at the hardware, operating system, middleware, or the application (algorithm) level. Synchronization is key to the correct coordination of the distributed algorithms. In general, there is no shared memory in a distributed system; therefore, all data transfers must be performed by *message passing* between the nodes.

1.5 Design Issues

Design issues and challenges in a distributed system may be broadly classified as in the area of system software and the distributed algorithms. Communication, synchronization, and the security problems are the key issues in the system software development side. Problems to be solved in distributed algorithms are numerous ranging from fault tolerance algorithms to load balancing to leader election in distributed systems. A distributed algorithm is designed to run at a node of a distributed system cooperating and synchronizing by other distributed algorithms running at other nodes of the distributed system to achieve a common goal. A *symmetric* distributed algorithm is executed on all nodes of the distributed system, whereas nodes may be running different components of an *asymmetric* distributed algorithm.

1.5.1 Synchronization

A fundamental problem in a distributed system is *time synchronization*, which aims at keeping the clocks of the nodes of the system in synchrony. As in a single processor system, access to shared resources must be monitored. In this so-called *mutual exclusion* problem, a number of algorithms were developed to provide mutual exclusion in distributed systems. *Deadlocks* in distributed systems may occur as in a single-processor system, where nodes of the distributed system wait for each other indefinitely, and no progress can be achieved. Precautions should be taken to prevent deadlocks. The analysis of distributed algorithms should provide proofs of deadlock-free executions. *Leader election* is another common problem where it is required to elect one of the all nodes or a group of nodes in the system to perform special tasks.

1.5.2 Load Balancing

It is a general requirement to distribute load that consists of code and data evenly to the nodes of the distributed system. The code and data of a process may need to be migrated from a heavily loaded node to a node with less load. The *response time*, which is the time taken from registering the input to providing a response to it, and *throughput*, which is the number of tasks finished in a given time, are two important metrics of performance in a distributed system. Load balancing aims to reduce the average response time and increase throughput in a distributed system.

While balancing the load, real-time requirements of the task should also be considered. A *hard real-time* task, such as a military application or a process control task, requires to be executed before a given deadline, and failure to do so may result in irreversible losses, whereas missing deadlines in a *soft real-time* system such as a banking system results in degraded performance.

1.5.3 Fault Tolerance

The aim of fault tolerance in distributed systems is to handle faults such as the crash of a computing node or a link connecting two nodes or a software module running at a node. Tolerance of faults is imperative in applications such as plant control or military applications. One way of achieving fault tolerance is by replicating code and data so that the replica may continue to work in the case of faults. The correct nodes reach agreement using *consensus algorithms*, which is another area of research in fault tolerant computing. *Check-pointing* and *recovery* procedures record the state of the software periodically on a secondary storage, and in case of faults, the system may be started from the last recorded state. These algorithms require significant synchronization in distributed systems.

Self-stabilizing algorithms aim at reaching a stable state in the presence of faults starting from any arbitrary initial condition. These algorithms should achieve a stable state in a bounded number of steps.

1.6 Distributed Graph Algorithms

The scope of the distributed algorithms in this book is confined to distributed graph algorithms, sometimes called *graph-theoretical distributed algorithms*, which exploit some property of the graph that represents the underlying communication network. For example, constructing a spanning tree of a graph is a well-studied problem, and there are few algorithms that find the spanning trees sequentially. Here, we will investigate how nodes of a distributed system cooperate to construct a spanning tree using their local knowledge of their neighbors.

The sequential graph algorithms are NP-Complete most of the time defying any solutions in polynomial time [4]. Using *heuristics* or *approximation algorithms* that find suboptimal solutions to the problems are the only choices in these situations. Heuristic approaches provide suboptimal solutions most of the time, but they do not guarantee these solutions. On the other hand, approximation algorithms guarantee to find a solution that approximates the optimal solution within a given factor. The task of the distributed graph algorithm designer then is twofold: to design an algorithm that is distributed and provide an approximation to the optimum solution at the same time.

The aim of this book is the design of such distributed approximation graph algorithms that may be of use in distributed applications. As a concrete example, finding a *minimum connected dominating set* that is the subset V' of vertices of a graph G with minimum size such that every vertex of the graph is either in V' or a neighbor of V' and all of the vertices in V' are connected is NP-hard for general graphs [5]. Therefore finding an approximation algorithm that has a better approximation than the best known algorithm is clearly a contribution on its own. Providing a distributed algorithm that approximates a connected dominating set either by modifying or improving the sequential solution or designing from scratch is also another contribution. A connected dominating set can be used as a backbone in an ad hoc wireless network. Modifying the distributed approximation algorithm now for an ad hoc wireless network by optimizing for energy levels and mobility of nodes is yet another challenge and may be a contribution on its own right. In summary, the contribution of the researcher in this field may be in few aspects; first, by designing an efficient approximation algorithm with a better approximation factor than the existing algorithms for the problem at hand; second, by providing a distributed version of the algorithm if this is possible and finally adapting this algorithm for ad hoc wireless networks by further introducing new parameters such as the mobility and energy levels of the nodes. Clearly, there are research challenges even in applying the well-established distributed approximation graph algorithms to ad hoc wireless networks.

1.7 Organization of the Book

Chapters in the book are organized in three parts. The first part describes fundamental graph algorithms starting by the construction of spanning trees in Chap. 4; graph traversal algorithms in Chap. 5; minimum spanning tree construction in Chap. 6; routing algorithms in Chap. 7; and self-stabilization in Chap. 8. Most of the algorithms in this part are well established, and our emphasis is on the implementation of these algorithms with detailed examples.

Part II is about graph-theoretical distributed approximation algorithms that mostly have applications in ad hoc wireless networks. These algorithms, as most of the algorithms provided in this book, use only local neighbor information most of the time and are called *local algorithms*. This part provides several recent algorithms with implementation details and examples. The algorithms are presented in an abstract manner without aiming at any specific application.

The algorithms developed in Parts I and II are reviewed and put into perspective for concrete network applications in Part III. This part starts by reviewing the model presented in Chap. 2, and we see that there have to be substantial changes. We also review some of the graph-theoretical algorithm concepts such as the dominating sets and provide new algorithms considering the additional parameters such as the mobility and energy level of the nodes in wireless ad hoc networks. Finally, a simple simulator that was developed to run distributed algorithms is presented with the implementation example to construct a spanning tree.

References

1. Akyildiz IF, Wang X, Wang W (2005) Wireless mesh networks: a survey. Comput Netw 47(4):445–487
2. Cokuslu D, Hameurlain A, Erciyes K (2010) Grid resource discovery based on centralized and hierarchical architectures. Int J Infonomics 3(1):227–233
3. Foster I, Kesselman C (2004) The grid: blueprint for a new computing infrastructure. Morgan Kaufmann, San Mateo
4. Garey MR, Johnson DS (1978) Computers and intractability: a guide to the theory of NP-completeness. Freeman, San Francisco
5. Guha S, Khuller S (1998) Approximation algorithms for connected dominating sets. Algorithmica 20(4):374–387
6. History of EGI. Homepage. http://www.egi.eu/about/EGI.eu/history_of_EGI.html
7. Mell P, Grance T (2011) The NIST definition of cloud computing. National Institute of Standards and Technology, US Dept. of Commerce, Special Publication, 800–145
8. Payli RP, Erciyes K, Dagdeviren O (2011) Cluster-based load balancing algorithms for grids. Int J Comput Netw Commun 3(5):253–269
9. US National Grid Homepage. http://www.fgdc.gov/usng

Part I
Fundamental Algorithms

Chapter 2
Graphs

Abstract Graphs are discrete structures that consist of vertices and edges connecting some of these vertices. Graphs have many applications in Mathematics, Computer Science, Engineering, Bioinformatics, and many other disciplines. Graphs are frequently used to model a communication network where computational nodes of a network are represented by vertices and the communication links between the nodes are represented by edges of the graph. In this chapter, we will review basic concepts in graph theory in relation to the modeling of a distributed system.

2.1 Definition of Graphs

Definition 2.1 (Graph) A graph is a tuple $G(V, E)$ where V is a nonempty set of *vertices* (or *nodes*) and E is a set of *edges*. Each edge has either one or two vertices as endpoints, that is, each edge is either a one- or two-element subset of V.

The vertex set V of a graph G may be infinite, in which case the graph is called an *infinite graph*, and a graph with a finite vertex set is called a *finite* graph. In this book, we will only consider finite graphs. For the graph $G = (V, E)$ and $v \in V$, the edge $e = \{v\}$ is called a *self-loop*. An edge is identified by the two vertices, and the edge is said to be incident to the vertices. For example, edge $e = \{v_1, v_2\}$, sometimes shown as $e = v_1 v_2$ or $e_{v_1 v_2}$, is incident to the vertices v_1 and v_2. The number of vertices of a graph ($|V|$) is called its *order*, and the number of its edges ($|E|$) is called its *size*. We will use literals n for the order and m for the size of a graph.

A graph that contains multiple edges connecting the same vertices is called a *multigraph*. A graph that does not contain edges that are self-loops and is not a multigraph is called a *simple graph*. We will only consider simple graphs in this book.

Definition 2.2 (Vertex Adjacency) Let $G(V, E)$ be a graph. Two vertices v_1 and v_2 are said to be *adjacent* if there exists an edge $e \in E$ that connects them so that $e = \{v_1, v_2\}$.

Definition 2.3 (Edge Adjacency) For a graph $G(V, E)$, two edges e_1 and e_2 are said to be adjacent if there exists a vertex v that is incident to (connects) both edges.

K. Erciyes, *Distributed Graph Algorithms for Computer Networks*,
Computer Communications and Networks, DOI 10.1007/978-1-4471-5173-9_2,
© Springer-Verlag London 2013

Fig. 2.1 (**a**) An undirected
simple graph; (**b**) a directed
multigraph

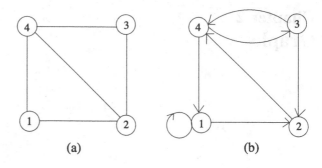

(a)   (b)

Based on these definitions, we can now define the neighborhood of a vertex as follows.

Definition 2.4 (Neighborhood) Given $G(V, E)$, the *neighborhood* of a vertex $v \in V$ is the set of vertices that are adjacent to v. Formally,

$$N(v) = \{u \in V : e(u, v) \in E\}.$$

$N(v)$ is usually called the *open neighborhood* of v, whereas $N[v] = N(v) \cup \{v\}$ is called the *closed neighborhood* of v, that is, the union of all neighbors of v and itself. The vertices of a graph are drawn as circles, and edges are the lines joining these vertices as shown in the example graph of Fig. 2.1(a), where $V = \{1, 2, 3, 4\}$ and $E = \{\{1, 2\}, \{2, 3\}, \{2, 4\}, \{3, 4\}, \{4, 1\}\}$. The neighborhood sets for vertex 2 are $N(2) = \{1, 3, 4\}$ and $N[2] = \{1, 2, 3, 4\}$. We will mostly use numbers to represent the vertices, unless this complicates description of an algorithm, in which case we will use letters.

Definition 2.5 (Degree) The degree of $v \in V$, deg(v), is the number of edges plus twice the number of self-loop edges incident to v.

The maximum degree of a graph is denoted by $\Delta(G)$, and the minimum degree by $\delta(G)$. $\Delta(G)$ of the graph in Fig. 2.1(a) is 3, and $\delta(G)$ is 2.

Up to now, we have considered *undirected graphs* that have *undirected edges*. However, in certain applications, such as the representation of data flow in computer networks, it may be required to assign directions to edges, in which case *directed graphs* are obtained.

Definition 2.6 (Directed Graph) A *directed graph* (*digraph*) $G(V, E)$ consists of a nonempty set of vertices V and a set of directed edges E where each $e \in E$ is associated with an ordered set of vertices.

An edge e that is associated with the ordered pair (u, v) is described as starting from u and ending at v. Figure 2.1(b) shows a digraph with $V = \{1, 2, 3, 4\}$ and $E = \{\{1, 1\}, \{1, 2\}, \{2, 4\}, \{3, 2\}, \{3, 4\}, \{4, 3\}, \{4, 1\}\}$.

Fig. 2.2 (a) A bipartite graph; (b) K_3; (c) K_4

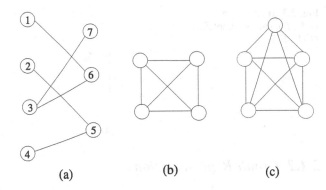

(a)   (b)   (c)

Definition 2.7 (In-Degree, Out-Degree) The *in-degree* of a vertex v in a digraph G is the total number of edges in E that end at v. The *out-degree* of v is the total number of edges in E that start from v. We will denote the in-degree of v by $\deg_{in}(v)$ and the out-degree by $\deg_{out}(v)$.

2.1.1 Special Graphs

We will describe some special graphs such as a *complete graph*, *bipartite graph*, and the *complement of a graph* in this part.

Definition 2.8 (Complete Graph) For the graph $G(V, E)$, if $\forall v \in V$, $N(v) = V \setminus \{v\}$, that is, if every vertex is connected to all other vertices of G, then G is called a *complete* graph. For a graph G with n vertices, the complete graph is denoted by K_n. For $K_n(V, E)$, $|E| = n(n-1)/2$.

Definition 2.9 (Bipartite Graphs) A graph $G(V, E)$ is called *bipartite* if V can be partitioned into two disjoint sets V_1 and V_2 such that every edge of G joins a vertex in V_1 to a vertex in V_2.

A bipartite graph with $V_1 = \{1, 2, 3, 4\}$ and $V_2 = \{5, 6, 7\}$ is shown in Fig. 2.2(a), and K_4 and K_5 are shown in Fig. 2.2(b) and (c).

Definition 2.10 (Complement of a Graph) The complement of a graph $G(V, E)$ is the graph $H(V, E')$ such that $e = \{v_1, v_2\} \in E'$ if and only if $e = \{v_1, v_2\} \notin E$. The complement of G is denoted G' or $\bar{G}$.

A graph G and its complement are shown in Fig. 2.3. A weighted graph $G(V, E, w)$ is a graph that has weights associated with edges, that is, $w : E \to \mathbb{R}$. Weighted graphs are frequently used to model communication networks as associated weights for edges may represent communication costs of sending messages over the links represented by the edges.

Fig. 2.3 (**a**) A graph
$G(V, E)$. (**b**) Its complement
$G'(V, E')$

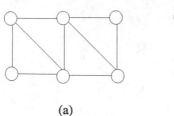
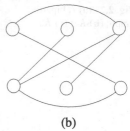

(a)   (b)

2.1.2 Graph Representations

In order to be able to perform some computation on graphs, they have to be represented in a format suitable for processing in a computer. Two important methods of representation are the *adjacency matrices* and *adjacency lists*.

Definition 2.11 (Adjacency Matrix) The *adjacency matrix* of a graph $G(V, E)$ with n vertices is an $n \times n$ matrix which has entry 1 at element (i, j) if there is an edge connecting vertex i to vertex j and 0 otherwise.

Definition 2.12 (Incidence Matrix) The *incidence matrix* of a graph $G(V, E)$ with n vertices and m edges is an $n \times m$ matrix which has entry 1 at element (i, j) if vertex i is incident to edge j and 0 otherwise.

Definition 2.13 (Adjacency List) The *adjacency list* of a graph $G(V, E)$ with n vertices is a list of n elements where each element consists of a vertex $v \in V$ and its neighbors connected using linked lists.

Figure 2.4 displays the adjacency matrix and the adjacency list of a graph.

2.2 Walks, Paths and Cycles

Definition 2.14 (Walk) A *walk* $w = (v_1, e_1, v_2, e_2, \ldots, v_n, e_n, v_n + 1)$ in G is an alternating sequence of vertices and edges in V and E, respectively, such that for all $i = 1, \ldots, n$, $\{v_i, v_{i+1}\} = e_i$. A walk is called *closed* if $v_1 = v_{n+1}$ and *open* otherwise.

Definition 2.15 (Trail, Tour) A *trail* in G is a walk in G where no edge is repeated and, a *tour* is a closed trail. An *Eulerian trail* is a trail that contains exactly one copy of each edge in E, and an *Eulerian tour* is a closed trail (tour) that contains exactly one copy of each edge

Definition 2.16 (Path) A *path* p from a vertex u to vertex v in graph G is a sequence of edges $e_1, \ldots, e_n$ such that each consecutive edge is incident to consecutive vertices along the path. The length of p is the number of edges it contains. When G

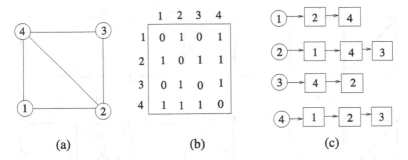

Fig. 2.4 (**a**) A graph $G(V, E)$. (**b**) Its adjacency matrix representation. (**c**) Its adjacency list representation

is simple, a path can be represented by the set of vertices $v_1, \ldots, v_n$ that it passes through (traverses). The path is called a *circuit* if it starts and ends at the same vertex. A *Hamiltonian Path* is a path that contains each vertex in V once. Alternatively, we can say that a path is a nontrivial walk with no edges and vertices repeated.

Definition 2.17 (Cycle) A *cycle* is a circuit of length of at least 3 and with no repeated edges except the first and last vertices. A *Hamiltonian cycle* is a cycle in a graph containing every vertex.

Definition 2.18 (Hamiltonian/Eulerian Graph) A graph $G = (V, E)$ is said to be *Hamiltonian* if it contains a Hamiltonian cycle and *Eulerian* if it contains an Eulerian tour.

A connected graph G is Eulerian if and only if every vertex of G has even degree. A connected graph G has Euler Trail if and only if the number of vertices with odd degree is less than or equal to 2. Figure 2.5 shows Hamiltonian Path, Hamiltonian Cycle, Eulerian Trail, and Eulerian Cycle. In (c), there are two odd-degree vertices as 2 and 8, and therefore an Eulerian Trail exists as shown. In (d), all vertices have even degrees, so an Eulerian Cycle exists as illustrated.

2.2.1 Diameter, Radius, Circumference, and Girth

Definition 2.19 (Distance) For a graph $G(V, E)$, the distance between the two vertices v_1 and v_2 in V is the length of the shortest walk beginning at v_1 and ending at v_2, provided that such a walk exists. We will write $d_G(v_1, v_2)$ to denote the distance between v_1 and v_2 in G.

Definition 2.20 (Diameter, Eccentricity, Radius) The *diameter* of G (diam(G)) is the length of the greatest distance in G. The *eccentricity* of v_1 is the maximum

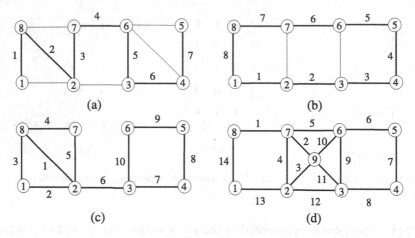

Fig. 2.5 (**a**) A Hamiltonian trail through vertices 1, 8, 2, 7, 6, 3, 4, 5. (**b**) A Hamiltonian path through vertices 1, 2, 3, 4, 5, 6, 7, 8, 1. (**c**) An Eulerian trail through vertices 8, 2, 1, 8, 7, 2, 3, 4, 5, 6, 3. (**d**) An Eulerian tour through vertices 8, 7, 9, 2, 7, 6, 5, 4, 3, 6, 9, 3, 2, 1, 8, all shown by *bold lines* and each edge labeled in sequence

distance from v_1 to any other vertex v_2 in V. The *radius* of G is the minimum eccentricity of vertices of G.

Definition 2.21 (Girth) For a graph $G(V, E)$, the *girth* of G is the length of the shortest cycle, provided that there is a cycle. When G does not have any cycle, the girth is defined as 0.

Definition 2.22 (Circumference) For a graph $G(V, E)$, the *circumference* of G is the length of the longest cycle, provided that there is a cycle in G. When G does not have any cycle, the circumference is defined as ∞.

The diameter of the graph in Fig. 2.5(a) is 4, for example, as the distance between vertices 1 and 5 through vertices 2–7–6. We will see that diam(G) is an important parameter in the determination of time complexities of distributed algorithms as it provides an upper bound on the time that a message is communicated between the two farthest points of a network graph.

2.3 Subgraphs

Certain applications may require finding solutions to a problem by computing the solution for small parts of the graph iteratively and then combining these partial solutions to obtain the final solutions. Informally, a smaller part of the graph is called a *subgraph*.

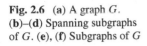

Fig. 2.6 (a) A graph G.
(b)–(d) Spanning subgraphs
of G. (e), (f) Subgraphs of G

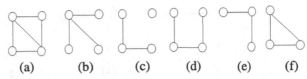

(a)　　(b)　　(c)　　(d)　　(e)　　(f)

Fig. 2.7 (a) A graph G.
(b) Edge-induced graph of G
of edges $\{1, 4\}$, $\{4, 3\}$.
(c) Vertex-induced graph of
G of vertices 2, 3, 4

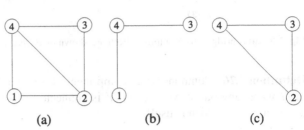

(a)　　　　(b)　　　　(c)

Definition 2.23 (Subgraph, Spanning Subgraph) A graph $H = (V', E')$ is called a *subgraph* of G if $V' \subseteq V$ and $E' \subseteq E$, with u and $v \in V'$; $\forall \{u, v\} \in E'$, that is, all vertices of H are also vertices of G and all edges of H are also edges of G. If $V' = V$, which means that H includes (covers) all vertices of G, then H is called a *spanning subgraph* of G.

Figure 2.6 shows the subgraphs of a graph.

Definition 2.24 (Edge-Induced Subgraph, Vertex-Induced Subgraph) Given an edge set $E' \subseteq E$, the edge induced subgraph by E' is $H = (V', E')$ where $v \in V'$ if and only if it is incident to an edge in E'. Similarly, given a vertex set $V' \subseteq V$, the vertex induced subgraph by V' is $H = (V', E')$ where $\{v_1, v_2\} \in E'$ if and only if both v_1 and v_2 are in V'.

Figure 2.7 shows the edge-induced and vertex-induced subgraphs of a graph.

2.4 Connectivity

An important property of a communication network is its capacity to withstand node and link failures. For example, it may be required to know the largest number of link failures that result in a disconnected network where there is no walk between every pair of computing nodes. Similarly, in graphs, we may need to determine the number of edge removals that will result in a disconnected network. Connectivity of a network is the determination of such parameters. Also, vertex and edge deletion methods are important in some of the algorithms that require removing a vertex from the graph at each iteration; we will see some of them in Part II.

Definition 2.25 (Connectedness) A graph $G(V, E)$ is *connected* if there is a walk between any pair of vertices v_1 and v_2. A digraph G is *strongly connected* if for every walk from every vertex $v_1 \in V$ to any vertex $v_2 \in V$, there is also a walk from v_2 to v_1 [3].

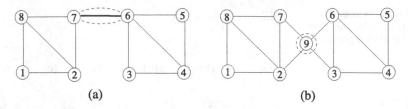

Fig. 2.8 (a) A bridge. (b) A cutpoint. Both are shown by *dashed lines*

Definition 2.26 (Component) A component of a graph $G(V, E)$ is a subgraph G' of G where any pair of vertices in G' is connected. A connected graph G has only one component which is itself.

Definition 2.27 (Edge Deletion Graph) For a graph $G(V, E)$ and $E' \subset E$, the graph G' formed after deleting the edges in E' from G is the subgraph induced by the edge set $E \setminus E'$, which is denoted $G' = G - E'$.

Definition 2.28 (Vertex Deletion Graph) For the graph $G(V, E)$ and $V' \subset V$, the graph G' formed after deleting the vertices in V' from G is the subgraph induced by the vertex set $V \setminus V'$, which is denoted $G' = G - V'$.

2.4.1 Cutpoints and Bridges

Definition 2.29 (Cutpoint) For a graph $G(V, E)$, a vertex $v \in V$ is a *cutpoint* of G if $G - v$ has more components than G has. If G is connected, $G - v$ is disconnected.

Definition 2.30 (Bridge, Cutset) For a graph $G(V, E)$, a *bridge* is an edge $e \in E$ deletion of which increases the number of components of G. A minimal set of edges whose deletion disconnects G is called a *cutset* in G.

The deletion of a bridge from a connected graph G provides two disconnected components of G.

Definition 2.31 (Block) A *block* of a graph G is its maximal subgraph that is connected and contains no cutpoints.

Figure 2.8 displays a bridge and a cutpoint of a graph. The subgraphs defined by vertices 1, 2, 7, 8 and 3, 4, 5, 6 are also blocks.

Definition 2.32 (Connectivity) The *vertex connectivity* (or just the *connectivity*) $\mathcal{K}$ of a graph G is the minimum number of vertices whose removal from G results in either a disconnected graph or a single vertex. The *edge connectivity* $\mathcal{E}(G)$ is defined as the minimum number of edges whose removal disconnects G.

2.5 Trees

Trees are important data structures in Computer Science as they have many applications such as database implementation, hereditary trees in bioinformatics, etc. A tree of a graph G also provides a graph with less edges and therefore with less communication links of the network. We will see many example algorithms to construct trees and implement distributed algorithms over the trees.

Definition 2.33 (Forest, Tree) A graph that contains no cycles is called *acyclic*. If $G = (V, E)$ is an acyclic graph and has more than one component, G is called a *forest*. If G has one component, then G is called a *tree*. Directed trees and forests are acyclic directed graphs.

The following are equivalent to describe a tree T:

- T is a tree;
- T contains no cycles and has $n - 1$ edges;
- T is connected and has $n - 1$ edges;
- T is connected, and each edge is a bridge;
- Any two vertices of T are connected by exactly one path;
- T contains no cycles, but the addition of any new edge creates exactly one cycle.

Definition 2.34 (Rooted Tree, parent, child, leaf) A tree is *rooted* if it has a designated vertex, called the *root*, in which case the edges have a natural orientation, toward or away from the root. In a rooted tree, the *parent* of a vertex is the vertex connected to it on the path to the root; every vertex except the root has a unique parent. A *child* of a vertex v is a vertex of which v is the parent. A *leaf* is a vertex without children.

Definition 2.35 (Spanning Forest, Spanning Tree) For graph $G(V, E)$, if $H(V', E')$ is an acyclic subgraph of G where $V' = V$, then H is called a *spanning forest* of G. If H has one component, it is called a *spanning tree* of G.

2.5.1 Minimum Spanning Trees

Definition 2.36 (Minimum Spanning Tree) For a weighted graph $G(V, E)$ where weights are associated with edges, a spanning tree H of G is called a *minimum spanning tree* of G if the total sum of the weights of its edges is minimal among all possible spanning trees of G.

If all weights of the edges of a graph G are distinct, then there is exactly one spanning tree of G. Figure 2.9 displays a possible spanning tree of a graph and its rooted minimum spanning tree.

Fig. 2.9 (a) A spanning tree.
(b) The minimum spanning
tree rooted at vertex 2

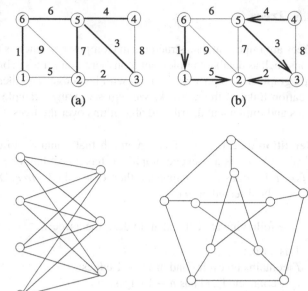

(a)   (b)

Fig. 2.10 (a) Complete
bipartite graph $K_{4,3}$.
(b) Petersen graph

(a)   (b)

2.6 Chapter Notes

The classical book on graph theory is by Harary [4] dating back to 1979. The text-book by West [5] is a more updated presentation of the topic. A thorough treatment of the topic is provided in the recent book by Bondy and Murty [1], and an informal presentation is provided in [2].

We have reviewed some of the basic concepts in graph theory. We will use undirected graphs to model the computer networks and develop distributed algorithms using this model. In Part III, we will need to modify this model to include ad hoc wireless networks.

2.6.1 Exercises

1. Show that the sum of the degrees of the vertices of an undirected graph is even. Show also that the number of odd degree vertices of an undirected graph is even.
2. For a bipartite graph $G(P, Q)$ where P and Q are disjoint vertex sets, show that $\sum_{u \in P} \deg(u) = \sum_{v \in Q} \deg(v)$.
3. A *degree sequence* of a graph G is the sequence of the degrees of the vertices of G in decreasing order. Find the degree sequences of the graphs in Fig. 2.8.
4. Show that for any graph G, $\text{rad}(G) \leq \text{diam}(G) \leq 2\,\text{rad}(G)$.

Fig. 2.11 An example graph
for Exercises 9 and 12

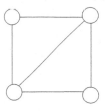

5. A simple graph G is called *regular* if all vertices of G have the same degree. In an *n-regular* graph G, all vertices have a degree of n. Determine the values of n for K_n and $K_{m,n}$ for these graphs to be n-regular.
6. Let G be a graph that has n vertices and m edges. Find the number of induced subgraphs and edge-induced subgraphs of G.
7. For which values of m and n does the complete bipartite graph $K_{m,n}$ have an Eulerian circuit and an Eulerian path?
8. Find the radius, girth, and diameter of the complete bipartite graph $K_{m,n}$ in terms of m and n and the Petersen graph shown in Fig. 2.10.
9. Draw all the subgraphs of the graph in Fig. 2.11.
10. Show that every tree with maximum degree k has at least k leaves.
11. A tree T with n vertices has a vertex of degree k. Prove that the longest path in T has at most $n - k + 1$ edges.
12. Find the spanning trees of the graph of Fig. 2.11.

References

1. Bondy JA, Murty USR (2008) Graph theory. Springer graduate texts in mathematics. Springer, Berlin. ISBN 978-1-84628-970-5
2. Fournier JC (2009) Graph theory and applications. Wiley, New York. ISBN 978-1-848321-070-7
3. Griffin C (2011) Graph theory. Penn State Math 485, Lecture notes. Homepage: http://www.personal.psu.edu/cxg286/Math485.pdf
4. Harary G (1979) Graph theory. Addison-Wesley, Reading
5. West DB (2001) Introduction to graph theory, 2nd edn. Prentice Hall, New York. ISBN 0-13-014400-2

Chapter 3
The Computational Model

Abstract In this chapter, we investigate how to model the application software, namely the distributed algorithm, the middleware, and the network that delivers the messages between the nodes of the distributed system.

3.1 Introduction

The computational model depends on the network model and the software environment that the distributed algorithm executes. As noted before, graphs are frequently used to model distributed systems. The vertex set V of a graph G represents the nodes of the network, and the edges show the communication links as shown in Fig. 3.1. A distributed algorithm runs at each node of the network graph and cooperates with other nodes to accomplish a common task. As an introductory example, let us attempt to design a simple routing algorithm for this network. In this network, node s wants to send a message $m(d)$ to node d. Nodes only know their neighbors, therefore, node i receiving $m(d)$ simply forwards this message to all of its neighbors, except the one it has received from, if the intended receiver d included in the header is not one of its neighbors.

Algorithm 3.1 displays the pseudocode for this algorithm for node i. It is assumed the a message is received from node j. If the network is connected, the message $m(d)$ will eventually reach the destination node d in at most *diam* time steps, where *diam* is the diameter of the network. As an example, the message sent by node 4 is flooded by receiving nodes until it reaches node 8, which knows that the destination 5 is its neighbor and sends the message to 5 only.

This algorithm has a major problem where a node may receive and then send the same message more than once, and the network may be flooded with duplicate messages. In order to remedy this situation, we could incorporate sequence numbers with the messages; therefore, each message carries the sender identifier i, destination identifier j, and a sequence number *seq* as $m(i, j, seq)$. Each node now can check whether it has seen the *seq* value from node i before. If it has, the message $m(i, j, seq)$ is a duplicate and can be discarded. But now, we need to store a table at each node to show the last received sequence number of message from each node. For a large network with n nodes, this table will be large. We have provided a method to overcome a problem but now faced a different problem. This situation is

K. Erciyes, *Distributed Graph Algorithms for Computer Networks*,
Computer Communications and Networks, DOI 10.1007/978-1-4471-5173-9_3,
© Springer-Verlag London 2013

Fig. 3.1 Simple routing
algorithm example

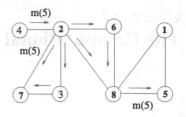

Algorithm 3.1 Simple Routing Algorithm

1: **int** i, j ▷ i is this node, j is the sender of the message
2: **message types** $m(sender, dest)$
3: **while** *true* **do**
4:     **receive** $m(j,d)$ ▷ receive message with destination d from neighbor j
5:     **if** $d \in \Gamma(i)$ **then**   ▷ if destination is a neighbor
6:         **send** $m(i, d)$ to d ▷ send message to the neighbor
7:     **else send** $m(i, d)$ to $\Gamma(i) \setminus \{j\}$ ▷ else send it to all neighbors except the sender
8:     **end if**
9: **end while**

---

not an exception; we may run into even more serious problems while trying to find
a solution to an existing problem while designing distributed algorithms. A simple
example has shown us that the contents of a message and where they are sent are
crucial in the design of distributed algorithms.

In this chapter, we will first analyze the steps in message delivery and how the
network behaves during this transfer in Sect. 3.2. We will then describe synchronization and the middleware primitives that provide the required coordination by
the application in Sects. 3.3 and 3.4, and then we will look into methods of spec-
ifying the coordination of the nodes from the view of the overall application in
Sect. 3.5. Finally, performance metrics of the distributed processing are described
in Sect. 3.6.

3.2 Message Passing

Messages are crucial for the correct operation of a distributed algorithm. We can
define the widely accepted *message passing model* of a distributed system formally
as follows [1, 3, 4]:

- A process p_i at node i communicates with other processes by exchanging mes-
  sages only.
- Each process p_i has a state $s_i \in S$, where S is the set of all its possible states.
- A configuration of a system consists of a vector of states as $C = [s_1, \ldots, s_n]$.
- The configuration of a system may be changed by either a *message delivery event*
  or a *computation event*.

Algorithm 3.2 Distributed Algorithm Structure

```
1: while condition do
2:     receive msg(j)
3:     case msg(j).type of
4:             type_A  :  Action_A
5:             type_B  :  Action_B
6:             type_C  :  Action_C
7: end while
```

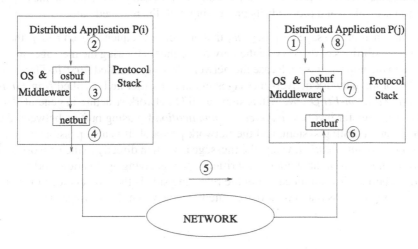

Fig. 3.2 Steps of message delivery

- A distributed system continuously goes through executions as $C_0, \phi_1, C_1, \phi_2, \ldots$, where ϕ_i is either a computation or a message delivery event.

A typical distributed algorithm code segment involves receiving a message and, based on the type of this message, performs a specific action as shown in Algorithm 3.2. Typically, the distributed algorithm runs until some condition is met, for example, a specific message is received, or a boolean variable becomes true.

The following steps are typically performed when the message $msg(j, type)$ is delivered from a distributed application process (algorithm) $P(i)$ at source node i to a distributed application process $P(j)$ at destination node j as shown in Fig. 3.2.

1. Receiving process $P(j)$ executes *receive(msg)* and is blocked by its local operating system at node j since there are no any messages.
2. Sending process $P(i)$ prepares the message by filling *data*, *destination process*, *node identifiers*, and *type* fields and invokes operating system primitive *send(msg, j)*, which copies *msg* to the operating system buffer *osbuf*.
3. The operating system copies *osbuf* to the network buffer *netbuf* and invokes the communication network protocol.

4. The protocol appends error checking and other control fields to the message and provides the delivery of the message to the destination node network protocol by writing contents of *netbuf* to the network link.
5. The network delivers the data packet, possibly by exchanging few messages and receiving acknowledgements.
6. The receiving network protocol at node j writes the network data to its buffer *netbuf* and signals this event to the operating system.
7. The receiving node's operating system copies data from *netbuf* to *osbuf* and unblocks the receiving process $P(j)$, which was blocked waiting for the message.
8. $P(j)$ is awaken and proceeds its processing with the received data.

If we consider messages m_1, m_2, m_3 that are sent in sequence from i to j, there are two possibilities of delivery by the network, either delivering the messages in sequence to the node j, in which case the network delivery is called *First-In-First-Out* (*FIFO*), or the network delivers messages in random order and is called *Non-First-In-First-Out* (*Non-FIFO*). We will assume a FIFO network structure in general. The above data transfer involved *buffered communication* by using buffers between the application, operating system, and the network protocol. It is also possible to have *unbuffered communication*, where the message is written directly to the network and received directly from network to be written to the receiving application workspace; however, buffered communication in the receiving side is the only choice mostly as the receiving process may not have executed *receive* when the message arrives.

3.3 Finite-State Machines

A finite-state machine (*FSM*) or finite-state automaton is a mathematical model to design systems whose output depends on the history of their inputs and their current states, in contrast to functional systems where the output is determined by the current input only. An FSM has a number of states, and it can only be at one state at any time called its *current state*. Upon a triggering by an event or a condition, an FSM may change its current state. States of an FSM are shown by circles, and the transitions from one state to another by directed edges. In this sense, an FSM is a directed graph. The edges of this graph are labeled as i/o, where i is the input in the form of an event or a condition, and o is the action performed by the FSM.

A deterministic FSM is a quintuple (I, S, S_0, δ, O) where

- I is a set of input signals
- S is a finite nonempty set of states
- $S_0 \in S$ is the initial start state
- δ is the state transition function such that $\delta : S \times I \rightarrow O$
- $O \in S$ is the set of output states

The state transition function decides on the next state using the current state and the input received. A state table shows the states as its rows and inputs as its columns, and the entries in the table are the actions or the next states. In a *Moore*

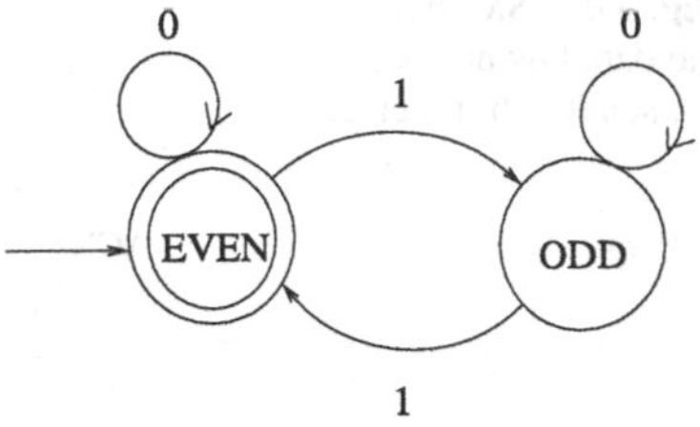

Fig. 3.3 FSM diagram of the
Parity Checker

Table 3.1 State table for the
Parity Checker

	0	1
ODD	ODD	EVEN
EVEN	EVEN	ODD

Machine type of FSM, the output is the new decided state. The *Mealy Machine* type
of FSM provides an output and a new state as a result of being triggered.

3.3.1 Moore Machine Example: Parity Checker

As an example of Moore Machine, let us design an FSM that inputs a binary string
of the form 001101010... and at any point, determines its state based on the number
of 1s received up to that point as ODD if this number is an odd number and EVEN
otherwise. This FSM called the *Parity Checker* has two states with the initial EVEN
state in double circles as shown in Fig. 3.3 and based on the binary input, it may
change its state. Since the output is equal to the next state this FSM decides, it is
a Moore Machine.

The FSM table for the Parity Checker has states ODD and EVEN shown as rows
and inputs 0 and 1, shown as columns in Table 3.1.

3.3.2 Mealy Machine Example: Data Link Protocol Design

As a more detailed example of a Mealy Machine FSM, we will consider the design
of a data link protocol called *Stop and Wait Automatic Repeat Request* (*ARQ*). The
main responsibilities of data link in general are the flow and error control of communication between the sender and the receiver network nodes. In the Stop and Wait
ARQ protocol, the sender has to wait for an acknowledgment from the receiver before sending the next frame. There will be a single frame in transmission at a time,
and to prevent the case where an acknowledgement from the receiver gets lost and
the sender sends a duplicate packet that is received as a new frame by the receiver,
odd and even sequence numbers are used.

Fig. 3.4 FSM diagrams of the data link protocol; **(a)** sender, **(b)** receiver

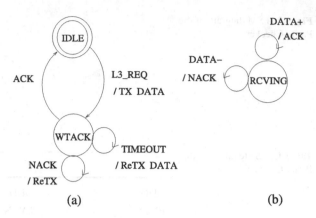

(a)   (b)

Table 3.2 State table for data link protocol sender

	L3_REQ	ACK	NACK	TIMEOUT
IDLE	Act_00	NA	NA	NA
WTACK	NA	Act_11	Act_12	Act_13

Figure 3.4 shows the FSM of the sender and the receiver processes of the protocol. The sender process initially is at IDLE state, and when there is a request from Layer 3 (Network Layer) to send a frame to the receiver, it prepares the frame header by inserting frame type INFO, sequence number, and error checking code there and transmits the frame after which it changes its state to WTACK to wait for a response from the receiver. In this state, it can receive either an *ack* frame from the receiver, which shows that the receiver has received the frame correctly, or a *nack* frame if the error checking at the receiver results in a negative result. If there are no replies from the receiver within a specified duration, the sender timeout expires, and it is notified by an interrupt, in which case it resends the frame. Transmission of the frame is also repeated when a *nack* is received.

The FSM table, Table 3.2, displays the states of the sender as rows and the possible inputs to the sender as columns. Any necessary action to be taken when input j is received at state i is placed as a procedure Act_ij in this table. This method of representing the FSM provides a convenient way of coding the FSM in a C-like language. Algorithm 3.3 shows a possible implementation where the addresses for all actions are placed in the table initially. The sender process keeps track of its current state, and when there is an input of any kind, it simply activates the action defined by its state and the input identifier. Using FSMs with this style of coding simplifies designing and implementation of complicated network protocols and distributed algorithms significantly.

We will be using FSMs to model some sample distributed algorithms to aid understanding them better, especially in cases where the algorithm is more complicated than usual.

Algorithm 3.3 Data Link Sender

```
 1: set of states IDLE, WTACK
 2: int currstate ← IDLE, curr_seqno ← 0
 3: fsm_table[2][3] ← action addresses

 4: procedure Act_00(frame)                                    ▷ send frame first time
 5:     frame.type ← DATA                                              ▷ set type
 6:     frame.seqno ← curr_seqno                            ▷ insert sequence number
 7:     frame.error ← calc_error(frame)              ▷ calculate and insert error code
 8:     send(frame) to receiver
 9:     currstate ← WTACK
10: end procedure
11:
12: procedure Act_11(frame)                              ▷ frame received correctly
13:     currseq ← (currseq + 1) mod 2                   ▷ increment sequence number
14:     respond to L3                                            ▷ notify Layer 3
15:     currstate ← IDLE
16: end procedure
17:
18: procedure Act_12(frame)                                     ▷ re-transmission
19:     if error_count ≤ MAX_ERR_COUNT then ▷ if maximum error count is not reached
20:         frame.type ← DATA                                          ▷ set type
21:         currseq ← (currseq − 1) mod 2              ▷ set seqno to the old one
22:         frame.seqno ← curr_seqno                     ▷ insert sequence number
23:         frame.error ← calc_error(frame)          ▷ calculate and insert error code
24:         send(frame) to receiver
25:     else send error_report to Layer 3         ▷ report delivery error to upper layer
26:     end if
27: end procedure
28:
29: while true do                                            ▷ Sender main code
30:     receive msg
31:     call fsm_table[currstate][msg.type] ▷ go to action specified by currstate and msg type
32: end while
```

3.4 Synchronization

Synchronization can be performed at various levels of a distributed system. At hardware level, processors may execute in lock step, and the next step of execution is not enabled until all nodes finish their current execution. This type of synchronization requires hardware support and is possible in *Single-Instruction-Multiple-Data SIMD* systems, where multiple processors perform the same computation on different data under a single control unit. *Multiple-Instruction-Multiple-Data (MIMD)* systems, however, do not rely on hardware synchronization, and nodes in such systems work autonomously. The MIMD systems represent distributed systems more realistically. Synchronization at network protocol level is accomplished by the send-

Fig. 3.5 Network layers

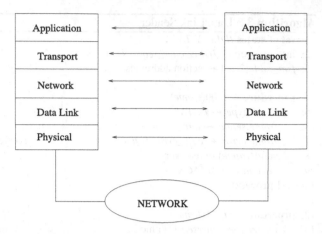

ing and receiving of messages to ensure correct delivery of a message to the destination. A typical network protocol at layer *l* communicates with layer *l* of the destination node by appending error checking codes to the message, receiving acknowledgements, and if there is an error, retransmission of the message is provided. Flow control and error-free delivery are the basic operations performed at various levels. Layer 2 (*Data Link Layer*) provides such functions, Layer 3 (*Network Layer*) of the protocol stack is responsible for the routing of the messages to the destination via optimal routes, and Layer 4 (*Transport Layer*) is responsible for delivery to the related application. The *Physical Layer* as the lowest layer provides all signal related functions and interface to the network. These protocol layers are displayed in Fig. 3.5.

There are also possibilities of synchronization at operating system level, middleware level, or the distributed application level. We will investigate these in the following sections.

3.5 Communication Primitives

Synchrony at operating system level is accomplished by carefully designing the communication primitives so that the sender and the receiver may be blocked or not by the operating system to yield the required behavior. The following are the possible send and receive primitives as supplied by the operating system:

- *blocking send*: The sending process is blocked by the operating system until an acknowledgement from the destination is received to confirm that the receiver has received the message, in which case the sender is unblocked.
- *nonblocking send*: The sending process continues processing after sending the message.
- *blocking receive*: The receiving process is blocked by the operating system if there are no messages available to it.

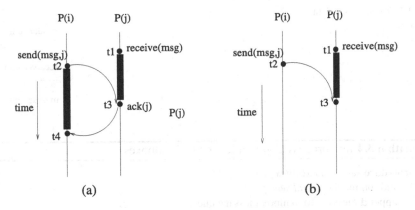

Fig. 3.6 (a) Blocking send and blocking receive. (b) Nonblocking send and blocking receive

- *nonblocking receive*: The receiving process checks if it has any pending messages, and if there is a message, receives it. In any case, it continues processing.

In Fig. 3.6(a), a fully synchronous communication is shown, where at time t_1, process $P(j)$ executes a blocking *receive*, and if there are no messages available at that time, it is blocked. At time t_2, process $P(i)$ invokes a blocking *send* that blocks itself and initiates the transfer of the message to node j, and the message is delivered to $P(j)$, which is awaken to continue. The first thing $P(j)$ does when awaken is sending the acknowledgement $ack(j)$ to the sending node i, which in turn unblocks the sender $P(i)$. In Fig. 3.6(b), however, the sender is not blocked assuming that the message is delivered to $P(j)$, which is unblocked when it receives the message. The nonblocking *receive* may be rarely used in situations where the receiver checks if there is a message but wants to continue even if there is not any message.

Blocking send is rarely used as correct delivery of the message is normally left to the network. Nonblocking receive requires specific storage in the operating system called *mailboxes*, which are used for indirect interprocess communications. A process may execute a nonblocking receive, which checks the mailbox and removes a deposited message. A mailbox is a depository place for a message, and writing a message to and receiving a message from a mailbox are provided by a data structure called *semaphore*, which consists of an integer and a process queue. There are two main operations on a semaphore, *wait* and *signal*. In a possible way of implementing a *wait* operation, the value of the semaphore is decremented, and if this is a negative number, the calling process is enqueued in the semaphore queue. The signal operation on a semaphore increments its value, and if this value is greater than or equal to zero, it dequeues and activates a waiting process from the semaphore queue. A mailbox data structure consists of a sender semaphore, a receiver semaphore, and a message queue as shown in Fig. 3.7.

A sending process to a mailbox has to check if there is space available in the mailbox; therefore, it performs a *wait* on the sender semaphore of the mailbox. It then deposits the message in the mailbox queue and signals the receiver semaphore of the mailbox to free any waiting process, if there is one, as shown in Algorithm 3.4.

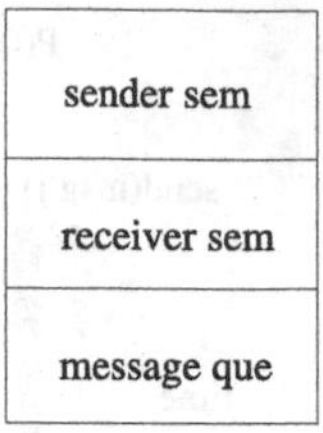

Fig. 3.7 The mailbox data
structure

sender sem
receiver sem
message que

Algorithm 3.4 Interprocess Communication by Mailboxes

 1: **procedure** *send_mbox*(*mbox_id*)
 2:     **wait** on mailbox *send semaphore*
 3:     **append** message to mailbox message queue
 4:     **signal** mailbox *receive semaphore*
 5: **end procedure**
 6:
 7: **procedure** *receive_mbox*(*mbox_id*)
 8:     **wait** on mailbox *receive semaphore*
 9:     **receive** message from mailbox message queue
10:     **signal** mailbox *send semaphore*
11: **end procedure**

In the case of a blocking receive, the receiving process simply performs a *wait* on
the receiving semaphore, and when this is successful, it proceeds by retrieving the
message from the mailbox queue. It also frees any waiting sender process by issuing
a *signal* on the sender semaphore of the mailbox. Using mailboxes for distributed
processing is possible by providing a naming facility to the operating system by
modifying the primitives as follows. The *send_mbox* routine checks the mailbox of
the receiver, and if this is not local, invokes the protocol stack routines. The receiving node protocol simply performs a local *send_mbox* to the mailbox of the receiver
on behalf of the remote sender as in [2].

3.6 Application Level Synchronization

We have investigated synchronization between pairs of application processes at operating system level. In this section, we will look into the synchronization among all
processes at the application layer, and we will call this synchronization the *global
synchronization*, which does not have any support from the operating system or the
hardware. In some applications, support from hardware or a special middleware
module called *synchronizer*, which provides synchrony among the processes, may
be a better choice than a synchronizing protocol as in the case of concurrently initiated synchronous algorithms. However, in many applications, synchronization at
network wide level may be achieved by the use of special protocol messages. In the

Table 3.3 Classification of distributed algorithms

	Single Initiator	Concurrent Initiator
Synchronous	SSI	SCI
Asynchronous	ASI	ACI

synchronous model, nodes of the network, and therefore the distributed applications on them, execute in a lock-step fashion in rounds as follows:

1. Start round
2. **send** message(s)
3. **receive** message(s)
4. perform computation

Typically, a distributed algorithm sends a message, receives a message, and performs some computation in each round. The next round can only be started after all messages from the previous round are delivered and all computations have been concluded. This seemingly strict control on the execution of processes usually requires the control of a central process, which starts a round, and after ensuring that the round is over, it starts the next round. We will see that this control can be achieved by a special message called *round* by a special node called the *root* that initiates a round, and after gathering special messages in an accumulated manner from all nodes, the root starts the next round.

In the asynchronous model, there is no restriction on the order of executions by the processes. However, we will assume that each message sent is correctly delivered to the application. Another distinction is whether the distributed application is started by a single designated process, in which case we will call the algorithm a *Single-Initiator Algorithm*, and when there are concurrent initiators, the algorithm is called a *Concurrent-Initiator Algorithm*.

A Synchronous Single-Initiator (*SSI*) algorithm is a synchronous distributed algorithm started by a single initiator. These algorithms are easier to analyze at the cost of requiring synchronous operation either at hardware level; at middleware level by a synchronizer; or by the addition of special protocol control messages. Asynchronous Single-Initiator (*ASI*) algorithms do not need any synchronization but should be designed carefully to provide termination condition so that the algorithm does not run forever. Synchronous Concurrent-Initiator (*SCI*) algorithms execute synchronously under the control of concurrent initiators and may require support from hardware such that stages of a round are performed after some n clock ticks by all processes in lock step manner. Asynchronous Concurrent-Initiator (*ACI*) algorithms are the most versatile of all but are more difficult to implement and analyze since synchronization of asynchronous algorithms at certain points during execution may be complicated. Table 3.3 shows the possible execution modes of distributed algorithms.

3.7 Performance Metrics

Performance of a distributed algorithm can be determined by its time, bit, space, and message complexities described below.

3.7.1 Time Complexity

For a sequential algorithm *Seq_Alg*, time complexity is the number of steps required for the algorithm to finish. We will show this parameter by Time(*Seq_Alg*). For a synchronous distributed algorithm *Synch_Dist*, Time(*Synch_Dist*) is the number of rounds required for the algorithm to finish in the worst case. For an asynchronous distributed algorithm *Asynch_Dist*, Time(*Asynch_Dist*) is the number of steps for the algorithm to finish in the worst case. For example, sending a message in a network modeled by a graph $G(V, E)$ may require $(n - 1)$ steps for it to reach the farthest node.

3.7.2 Bit Complexity

For a distributed algorithm, we will mostly be interested in the maximum length of a message communicated. In general, this will not be a problem unless the message is large or is enlarged as it traverses the network. For example, an application may require that a special message should include the nodes identifiers in it, as it visits the nodes of the network in some order. In this case, message will include a maximum of n node identifiers. We will assume that $\log n$ bits are necessary to hold a node identifier, and the bit complexity for this example algorithm will then be $O(n \log n)$.

3.7.3 Space Complexity

Space complexity of an algorithm specifies the maximum storage in bits required by the algorithm for local storage at a node. This may be important if a node holds large tables as in the case of routing algorithms.

3.7.4 Message Complexity

Number of messages exchanged is a good indicator of the cost of communication for the distributed application. The cost incurred during the message communication of a distributed algorithm is often considered as the dominant cost of the algorithm

---

Algorithm 3.5 Sample_SSI

1: **boolean** *finished, round_over* ← *false*
2: **message type** *round, info, upcast*
3: **while** ¬*round_over* **do**
4:      **receive** *msg(j)*
5:      **case** *msg(j).type* **of**
6:                    <u>*round*</u>:   **send** *info(i)* to all neighbors
7:                                **receive** *info* from all neighbors
8:                                **do** some computation, *finished* ← *true*
9:              <u>*upcast*</u>:   **if** *upcast* received from all children **and** *finished* **then**
10:                               **send** *upcast* to *parent*
11:                               *round_over* ← *true*, *finished* ← *false*
12: **end while**

---

since time spent in message transmissions is orders of magnitude higher than the time spent for local computations. A convenient method of evaluating communication costs is to calculate the number of messages that traverse the edges of the network graph. Let us demonstrate this concept by an SSI algorithm, a single round of which is shown in Algorithm 3.5, where a spanning tree T is formed prior to the algorithm execution such that every node except the root has a parent and any node other than the leaves have children. T will be used to transfer the synchronization messages *round* and *upcast*. The *root* initiates each round by the *round* message, and the *upcast* messages are gathered by the nodes from their children to send to their parents to end a round. A node sends an *upcast* message to its parent only when it has finished computation and all of its children have also finished computation and have sent *upcast* messages to it. The flag *finished* is needed as a node may receive *upcast* messages from its children before it can finish its computation.

Each node in each round sends *info* messages to all its neighbors, receives *info* messages from all of its neighbors, and does some computation. In order to calculate the number of messages exchanged in *Sample_SSI*, we start by counting the messages exchanged at each round. The total number of messages for exchanging information in each round is $2|E|$ as each edge of the network graph is traversed exactly twice, once in each direction by *info* messages. There will also be $n - 1$ *round* messages and $n - 1$ *upcast* messages as T has $n - 1$ edges, for a total of $2n - 2$ messages for synchronization. Total number of messages in each round therefore is $2(m + n - 1)$. The number of rounds would depend on the application, and if we assume that the requirement for *Sample_SSI* is the transfer of the local node knowledge to all nodes in the network, then the number of rounds will be the diameter d, which is the number of hops between the two farthest nodes of the network, and d could be as high as $n - 1$. The total number of messages for *Sample_SSI* then is $2d(m + n - 1)$. We will see more examples of calculating the total number of messages for different applications. As for the time complexity of *Sample_SSI*, assuming that T may have $n - 1$ as its maximum depth, there will be $n - 1$ time steps in each round to deliver *round* messages to all nodes; each node will require two time units to send *info* message to neighbors and receive *info* message from the neighbors; and there will be further $n - 1$ steps for the accumulation of *upcast* messages for a total of $2n$ steps in each round resulting in a total number of at most $2n(n - 1)$ time steps for $n - 1$ rounds. However, the total number of rounds is a good indicator and sufficient for the time complexity of an SSI algorithm in general, in which case Time(*Sample*_SSI) = $O(d)$ for the sample algorithm.

3.8 Chapter Notes

The computational model for the distributed application has few aspects to be considered, with the synchronization being the key issue to be addressed. Synchronization can be handled at one or more of hardware, operating system, middleware, and application levels. Synchronization at hardware level may provide system-wide synchronization but is difficult due to the distributed nature of the application. Operating system may provide synchronization between pairs of processes but will not provide system-wide synchronization unless augmented by special software called synchronizers. Global synchronization may be achieved by the use of special graph structures such as a spanning tree of the graph to convey special messages to start and finish each round. However, even this seemingly versatile method may have significant overhead if the application requires synchronization in a concurrently initiated algorithm.

Another aspect of the model is concerned with how the algorithm is initiated. We have seen the four possible cases of SSI, ASI, CSI, and CAI. SSI may be a method preferred in many cases as it simplifies the design significantly; ASI and CAI are also used in various applications, but CSI is probably the last choice due to its special requirements outlined above. We have also described how modeling by FSMs simplifies the design and implementation of the distributed algorithms and network protocols. We will use the SSI model frequently, with the control messages transferred over a spanning tree to implement various distributed algorithms.

3.8.1 Exercises

1. Provide a pseudocode for the nonblocking receive primitive with brief comments.
2. The Front End Process (FEP) of a file manager in a distributed system inputs messages and, based on the requests that can either be *read* from, *write* to, or *copy* a file, invokes the necessary process by sending a mail in its mailbox. The FEP returns to wait for any more incoming messages while the related process performs the required action. Describe a possible message frame format for the protocol manager and write pseudocodes for these four processes with brief comments.

3. It is required to have a general *send* procedure that checks whether the receiver is local and, if not, invokes the protocol manager by depositing the message in its mailbox. Provide the pseudocode for this procedure with brief comments
4. An SSI distributed algorithm is executed by a root node over an already formed spanning tree to find the largest degree of the network graph. The root sends a *probe* message to its children, which is transferred to the leaves. Provide pseudocodes for the root and ordinary nodes.
5. An SSI distributed algorithm executed at a node of a computer network aims at finding the largest identifier node within two-hop distances from each node.
   a. Provide a pseudocode for a single round of this algorithm.
   b. Work out its time and message complexities.

References

1. Attiya H, Welch J (2004) Distributed computing: fundamentals, simulations, and advanced topics, 2nd edn. Wiley, New York
2. Erciyes K (1989) The design and implementation of a real-time multitasking kernel for a distributed operating system. Ph.D. thesis, Ege University
3. Lynch NA (1996) Distributed algorithms. Morgan Kaufmann, San Mateo
4. Tel G (2000) Introduction to distributed algorithms, 2nd edn. Cambridge University Press, Cambridge

Chapter 4
Spanning Tree Construction

Abstract Spanning trees have many applications in computer networks as they provide a subgraph of less number of links than the original network resulting in lowered communications. This chapter introduces the basic distributed algorithms to construct spanning trees of graphs without any particular optimization objective.

4.1 Introduction

A *spanning tree* of a connected, undirected graph $G(V, E)$ is its subgraph $T(V, E')$ that covers (spans) all vertices of G. A spanning tree of a connected graph G can also be defined as a maximal set of edges of G that contains no cycle, or as a minimal set of edges that connect all vertices. A *spanning forest* of G is a subgraph of G that consists of a spanning tree in each connected component of G.

Spanning trees are important structures for computer networks as they provide a subgraph of G with possibly less communication links, resulting in lowered communication costs. Providing a parent/child relationship among nodes of the network eases the task of communication since the source and destination of communication is known beforehand. In this chapter, various distributed algorithms without any optimization objective to construct spanning trees for computer networks are presented with added complexity at each stage. First, the *Flood* algorithm, which assumes no prior structure of the network, is introduced in Sect. 4.2 to provide a broadcast of a message from a root process to all nodes of the network. Then, using *Flood*, forming a spanning tree that can be used for further broadcasts or other communication tasks is described in Sect. 4.3. The algorithm described in Sect. 4.4 provides detection of the termination of the algorithm by the nodes in the network utilizing a collection of acknowledgement messages starting from the leaf nodes. Tarry's algorithm, which builds a spanning tree by using two simple rules, is shown in Sect. 4.5. We conclude this chapter by showing how efficient broadcast and convergecast operations can be achieved using the constructed spanning trees.

4.2 The Flooding Algorithm

Many applications in computer networks require sending a message to all nodes in the network that is called the *broadcast*. A natural way of performing broadcast in a

K. Erciyes, *Distributed Graph Algorithms for Computer Networks*,
Computer Communications and Networks, DOI 10.1007/978-1-4471-5173-9_4,
© Springer-Verlag London 2013

Algorithm 4.1 *Flood*

1: **int** i, j
2: **boolean** *visited* $\leftarrow$ *false*
3: **message types** *msg*
4:
5: **if** $i = root$ **then**   ▷ *root* initiates flooding
6:     **send** *msg* to $\Gamma(i)$
7:     *visited* $\leftarrow$ *true*
8: **end if**
9:
10: **receive** *flood*(j)   ▷ *flood* may be received many times
11: **if** *visited* $=$ *false* **then**   ▷ *msg* received first time
12:     **send** *msg* to $\Gamma(i) \setminus \{j\}$
13:     *visited* $\leftarrow$ *true*
14: **else**   ▷ *msg* received before
15:     **discard** *msg*
16: **end if**

---

network without any formed structure is to simply forward any incoming message to all neighbor nodes except the neighbor that has sent the message. If the same message arrives again, it should be discarded. The algorithm *Flood*, which performs broadcast in this simple form, initiated by a specific node called the *root*, is shown in Algorithm 4.1 for node i.

4.2.1 Analysis

Theorem 4.1 *The message complexity of Flood is $O(m)$ where m is the number of edges of G, and the time complexity of Flood is $\Theta(d)$ where d is the diameter of G.*

Proof Since each edge connects two nodes and is used to deliver a message at least once and at most twice when two nodes send *msg* concurrently, there will be a total of $2m$ messages at most, and therefore, Msg(*Flood*) $= O(m)$. The longest time for the broadcast message to reach any node in the graph G is the distance between two farthest nodes of the graph, which is the diameter, and hence, Time(*Flood*) $= \Theta(d)$. $\square$

It can be easily seen that this algorithm is inefficient as each edge of G may be utilized more than once. An improvement can be achieved if the flooding algorithm can be used to build a spanning tree rooted at the initiator, and this spanning tree may then be used for any further broadcast messages, as described in the next section.

Algorithm 4.2 *Flood_ST*

```
 1: int parent ←⊥
 2: set of int childs ← ∅, others ← ∅
 3: message types probe, ack, reject
 4:
 5: if i = root then                                    ▷ root initiates tree construction
 6:     send probe to Γ(i)
 7:     parent ← i
 8: end if
 9:
10: while (childs ∪ others) ≠ (Γ(i)\{parent}) do
11:     receive msg(j)
12:     case msg(j).type of
13:             probe:  if parent = ⊥ then               ▷ probe received first time
14:                         parent ← j
15:                         send ack to j
16:                         send probe to Γ(i)\{j}
17:                     else                             ▷ probe received before
18:                         send reject to j
19:             ack:    childs ← childs ∪ {j}                ▷ include j in children
20:             reject: others ← others ∪ {j}     ▷ include j in unrelated neighbors
21: end while
```

4.3 Flooding-Based Asynchronous Spanning Tree Construction

We can use the algorithm *Flood* by some modifications to build a spanning tree originating from the initiator *root* for broadcasting. We assume that it is required that each node in the tree except the leaf nodes should know the identifiers of its children and all nodes except the root should know their parents in the end. Any node that wants to build a broadcast tree initiates the algorithm and becomes the *root* of the spanning tree to be formed.

The messages used in this algorithm, called algorithm *Flood_ST*, are *probe*, *ack*, and *reject*. Any node that wants to build a spanning tree starts the algorithm by sending the *probe* message to its neighbors, which is transferred to other nodes. Since a node may receive more than one *probe* message, acknowledgement (*ack*) and negative acknowledgement (*reject*) messages are needed to check whether a node has received *probe* before as shown in Algorithm 4.2.

The *root* starts the algorithm, and whenever node *i* receives a *probe* message, it marks the sender *j* as its parent and sends an *ack* message to *j*. The parent *j*, in receipt of an *ack*, marks the child as one of its children. Then, node *i* sends *probe* message to all of its neighbors except the parent *j* consequently. If a node already has a parent when it receives a *probe* message from a neighbor node, it sends a *reject* message to the neighbor. The termination condition is when the union of the children (*childs*) and unrelated neighbors (*others*) of a node *i* equals its neighbors except the parent, as checked in line 10 of the algorithm. It should be noted that the main body of the algorithm between lines 10–21 is also executed by the *root*.

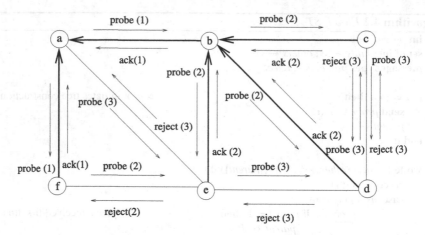

Fig. 4.1 An example spanning tree formed by *Flood_ST* algorithm

Figure 4.1 shows an example spanning tree formed by the *Flood_ST* algorithm over a network consisting of six nodes a, b, c, d, e, and f. Node a is the root and starts construction of the tree by sending *probe* messages to its neighbors b, e, and f. Each message is labeled with the time frame number such that messages with the same label occur concurrently within the same time frame. It can be seen that due to the delay in the link between a and e, the probe message from b reaches e before, and b becomes the parent of the node e although it is now two hops away from the root. The final constructed spanning tree is shown by the bold lines. The total time taken is thee units as shown by the labels of the last communicated *ack* and *reject* messages. The total number of messages is 20, which is $2m + 2$, m being 9 for this network, as there are two concurrent *probe* messages sent between the neighbors c and d, both of which are rejected resulting in two extra traversals of the edge $\{c, d\}$.

4.3.1 Analysis

Theorem 4.2 *The message complexity of algorithm Flood_ST is $O(m)$ where m is the number of edges of G, and it builds a tree T of maximum depth $n - 1$. Assuming that there is at least one message transfer at each time unit, its time complexity is $O(n)$.*

Proof Each edge of G will be traversed at least twice with *probe* and *ack*, or with *probe* and *reject* messages, or at most four times in the case of two nodes attempting to send each other *probe* messages concurrently. They will both reply with *reject* messages resulting in four messages for this edge for a total of $4m$ messages. Therefore, Msg(*Flood_ST*) = $O(m)$. The messages may be transferred over the fast communication links of the longest path instead of the shorter paths with slow links as

Fig. 4.2 An example of the longest path

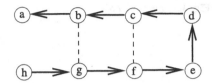

shown in Fig. 4.2. Node a starts the algorithm in this network, and assuming that the links between nodes b and g and between c and f are much slower than the other links, the longest path between the two farthest nodes a and h is taken, which has a length of 7, which is $n - 1$ as n is 8 in this network. The depth of the tree constructed is $O(n)$ considering the longest path. Assuming that there is at least one message transaction at each time step, the time complexity is bounded by the longest path in the graph, which has a length of $n - 1$.   □

The problem with the *Flood_ST* algorithm is that although the root and nodes determine that their part of algorithm is over, they are not aware that the algorithm has terminated globally. The algorithm described in the next section provides the necessary modification to *Flood_ST* so that the nodes know when the algorithm has finished at least in their subtrees.

4.4 An Asynchronous Algorithm with Termination Detection

As a further attempt to build a spanning tree asynchronously, we will modify the previous algorithm so that termination of the construction is detected by the nodes. The modification is achieved by the nodes delaying the sending of the *ack* message to their parents until they receive *ack* or *reject* messages from their neighbors, rather than replying to their parents immediately. This way, when a node receives all the replies from its neighbors, it can determine that all of the nodes in the subtree in which it is the root has terminated. We will describe this algorithm, called *Term_ST*, by using a finite-state machine (FSM).

The FSM of this algorithm is shown in Fig. 4.3. All nodes start from the IDLE state, and at the end of the algorithm, every node should finish in terminated (TERM) state. The initiating *root* node starts the algorithm by sending a *probe* message to its neighbors in $\Gamma(i)$. Node i receiving the *probe* message from a neighbor node j for the first time, enters the explored (XPLORD) state and marks the sender as its parent. The parent, in return, marks the child as one of its children. If a node already has a parent when it receives a *probe* message from a neighbor node, it sends a *reject* message to the neighbor. Differently from the *Flood_ST* algorithm, node i now defers sending of an *ack* message to its parent until it receives *ack* and *reject* messages from all of its neighbors except the parent. It does however send *probe* messages to its neighbors except the parent immediately. Only a *leaf* node that does not have any children and that receives *reject* messages from all of its neighbors except the parent would initiate the termination of the algorithm as

Fig. 4.3 *Term_ST FSM*

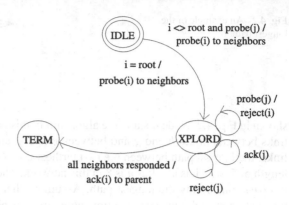

Algorithm 4.3 *Term_ST*

1: **int** *currstate* ← *IDLE*, *parent* ←⊥
2: **set of int** *childs* ← ∅, *others* ← ∅
3: **message types** *probe*, *ack*, *reject*
4: **if** *i* = root **then**
5:     **send** *probe* to *Γ(i)*
6:     *currstate* ← *XPLORD*
7: **end if**
8:
9: **while** (*childs* ∪ *others*) ≠ (*Γ(i)\{parent}*) **do**
10:     **receive**(*msg(j)*);
11:     **case** *currstate* **of**
12:         *IDLE*:
13:             **case** *msg(j).type* **of**
14:                 *probe*:   *parent* ← *j*   ▷ *probe* received first time
15:                              **send** *probe* to *Γ(i)\{j}*
16:                              *currstate* ← *XPLORD*
17:         *XPLORD*:
18:             **case** *msg(j).type* **of**
19:                 *probe*:   **send** *reject* to *j*   ▷ *probe* received before
20:                 *ack*:   *childs* ← *childs* ∪ {*j*}
21:                 *reject*:   *others* ← *others* ∪ {*j*}
22: **end while**
23: **if** *i* ≠ *root* **then**   ▷ convergecast *ack* to root
24:     **send** *ack(i)* to *parent*
25: **end if**
26: *currstate* ← *TERM*

---

shown in Algorithm 4.3. The termination condition is when the total set of nodes
that have responded is equal to the union of the children and unrelated nodes as
shown in line 9 of the algorithm.

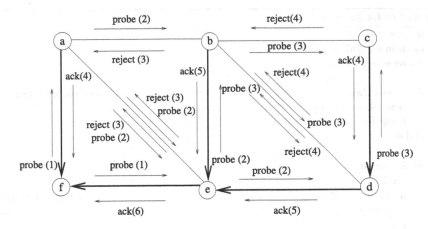

Fig. 4.4 An example operation of *Term_ST*

4.4.1 Analysis

Theorem 4.3 *The message complexity of algorithm Term_ST is $O(m)$, and assuming that there is at least one message transfer at each time unit, its time complexity is $O(n)$.*

Proof The message complexity can be determined as in the *Flood_ST* algorithm to result in $O(m)$. Due to the asynchronous operation, messages may take the longest path of length $n-1$ to form the tree, and since there will be $n-1$ more steps for the reply messages to be gathered at the root along this longest path, there will be a total of $4n-2$ time steps at most, considering there is at least one message transfer at each time unit. Time(*Term_ST*) is therefore $O(n)$. Otherwise, the time to construct a spanning tree is unbounded. □

Figure 4.4 shows an example operation of *Term_ST* where node f is the root and starts the algorithm by sending the *probe* message to its neighbors a and e, which in turn mark node f as their parent and send *probe* messages to their neighbors b, e and a, b, d, respectively. Node b receives *probe* from e first, marks e as the parent, and sends *reject* to node a in time frame 3 when *probe* from a arrives as it is now in *XPLORD* state. At this point, node a has received responses from all of its neighbors and now sends *ack* to parent f and terminates. The operation continues similarly until all *ack* messages are cast to the root f. Time taken for the construction of the spanning tree is 6 units, assuming that a node responses to the probe message in the same time interval. As m is 9 for this network, there are $2m + 4 = 22$ messages in total since each edge is traversed twice except edges $\{a, e\}$ and $\{b, d\}$, which are traversed additional two times due to the concurrent sending of *probe* messages.

In order to provide concurrent initiators to form spanning trees rooted at each initiator, the *Term_ST* algorithm can be modified so that each node has n states, and

---

Algorithm 4.4 *Tarry_ST*

1: **int** *parent* ← ⊥
2: **boolean** *used*[*n*] ← {*false*}
3: **message types** *token*
4:
5: **if** *i* = *root* **then**   ▷ root starts the search
6:     **send** *token*(*i*) to any *j* ∈ Γ(*i*)
7:     *used*[*j*] ← *true*, *parent* ← *i*
8: **end if**
9:
10: **while** *true* **do**
11:     **receive** *token*(*j*)
12:     **if** *parent* = ⊥ **then**   ▷ token first time
13:         *parent* ← *j*
14:     **end if**
15:     **if** ∃*j* ∈ (Γ(*i*) \ *parent*) : ¬*used*[*j*] **then**   ▷ choose an unsearched neighbor ≠ parent
16:         **send** *token* to *j*
17:         *used*[*j*] ← *true*
18:     **else**
19:         **if** *i* ≠ *root* **then**
20:             *used*[parent] ← *true*   ▷ all neighbors searched
21:             **send** *token* to *parent*
22:         **end if**
23:         **exit**   ▷ terminate
24:     **end if**
25: **end while**

---

the identity of each initiator is tagged in the message so that each node will have a different position in *n* spanning trees (see Exercise 5).

4.5 Tarry's Traversal Algorithm

Tarry's algorithm *Tarry_ST* is a very early distributed algorithm that builds a spanning tree by the traversal of a token using two simple rules shown below [4]. There is a designated *root* node as before, and when node *i* receives the *token* for the first time from node *j*, it marks it as its parent. Since only a node that has the token is enabled, there is a single point of activity at any time.

1. A process never forwards the token twice through the same channel.
2. A noninitiator forwards the token to its parent, the node from which it received the token for the first time, only if there is no other channel left according to Rule 1.

In order to implement Rule 1, node *i* uses an array *used* to monitor status of its neighbors. Upon reception of the token, node *i* forwards the token to an unsearched neighbor *j*, assigns true value to *used*[*j*], and when all of the neighbors have true

Fig. 4.5 Tarry's algorithm operation

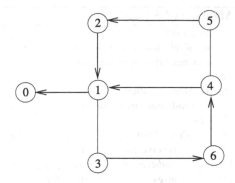

values in the *used* array meaning that all neighbors have been searched, it forwards the token back to its parent to implement Rule 2. The pseudocode for the algorithm is shown in Algorithm 4.4.

Figure 4.5 shows an example execution of *Tarry_ST* algorithm in a network of 7 nodes. A possible traversal of token in this network is 0 1 4 5 2 1 3 6 4 6 3 1 2 5 4 1 0.

4.5.1 Analysis

Theorem 4.4 *The time complexity of Tarry_ST is* $\Theta(m)$, *and its message complexity is also* $\Theta(m)$.

Proof Each edge is used to deliver a message exactly twice, once in each direction governed by the rules for a total of $2m$ times. Since there is a single point of activity at any time, there will be $2m$ steps, and hence Time($Tarry\_ST$) $= \Theta(m)$ and Msg($Tarry\_ST$) $= \Theta(m)$. □

4.6 Convergecast and Broadcast over a Spanning Tree

In a computer network, it may be required to gather data from all nodes of the network to the *root* node of an already formed spanning tree. This operation, called the *convergecast*, is one of the key data transfer operations in a wireless sensor network. Algorithm 4.5 provides asynchronous gathering of data starting from the leaves of an already formed spanning tree. The key to the operation of this algorithm is that any nonleaf node should wait data from all of its children before uploading the combined/processed data to its parent. The convergecast operation also allows manipulation of data received from all of the children of a node such as calculating the average value of received data from all of the children and then sending this to the parent. The algorithm ends when the root receives the convergecast messages from all of its children.

Algorithm 4.5 *Ccast_ST*

```
 1: int parent
 2: set of int childs; gathered ← ∅
 3: message types convcast; msgs ← ∅
 4:
 5: if childs = ∅ then                                  ▷ leaf nodes start convergecast
 6:     send convcast to parent
 7: else                                                ▷ any intermediate node or root
 8:     while childs ≠ gathered do      ▷ wait for convergecast messages from all children
 9:         receive convcast( j)
10:         gathered ← gathered ∪ { j}
11:         msgs ← msgs ∪ convcast( j)
12:     end while
13: end if
14: if i ≠ root then
15:     combine msgs into convcast
16:     send convcast to parent
17: end if
```

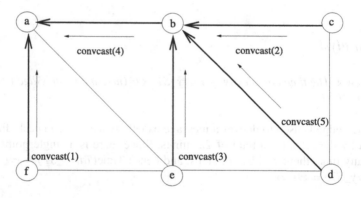

Fig. 4.6 Convergecast operation

Figure 4.6 shows the operation of the algorithm *Ccast_ST* on an example spanning tree. Assuming that messages are labeled by their duration of transmission and all leaves start convergecast concurrently, the total time for the convergecast at node *b* is 5 units, and convergecast from node *f* to *a* is also performed in this period. Convergecast from node *b* to *a* takes a further 4 units, and the total duration for the convergecast of all data at node *a* is 9 units.

Broadcasting a message *m* over a spanning tree *T* is initiated by the *root* node, which sends *m* to all its children and is terminated when all nodes receive *m*. Any node that receives *m* simply forwards it to all its children, and a leaf node does not forward *m*. Algorithm 4.6 shows the broadcast operation that is augmented by an inline convergecast operation to enable the root know when the broadcast operation has terminated. A *leaf* node starts the convergecast operation when it receives the

---

Algorithm 4.6 *Bcast_ST*

1: **set of int** *childs* ← ∅, *acked* ← ∅
2: **message types** *bcast*, *ack*
3: **if** *i* = *root* **then**
4:     **send** *bcast* to *childs*
5: **end if**
6: **while** *acked* ≠ *childs* **do**   ▷ collect *acks* from *childs*
7:     **receive** *msg(j)*
8:     **case** *msg(j).type* **of**
9:          *bcast*:   **if** *childs* ≠ ∅ **then**
10:                        **send** *bcast* to *childs*
11:                        **else send** *ack* to *parent*   ▷ start convergecast
12:          *ack*:   *acked* ← *acked* ∪ {*j*}
13: **end while**
14: **if** *i* ≠ *root* **then**
15:     **send** *ack* to *parent*
16: **end if**

---

broadcast message. When the root receives convergecast messages from all of its children, it can determine that all nodes have received the broadcast message, and it terminates. The depth of T may be as large as $n - 1$, and for this reason, forming a breadth-first-search tree for ordinary broadcast is more efficient as described in the next chapter.

Theorem 4.5 *The message complexity of both Ccast_ST and ordinary broadcast algorithm Bcast_ST is $\Theta(n)$. The time complexity of both algorithms is $\Theta(depth(T))$, which would be at most $n - 1$.*

Proof In both algorithms, each edge of T is used to deliver a message once, and since the total number of edges of an n node tree is $n - 1$, there will be a total of $n - 1$ messages. For the *Bcast_ST* algorithm described, which provides a convergecast operation, there will be a further $n - 1$ *ack* messages convergecast to the root, resulting in a total of $2n - 2$ messages. Time for broadcast or convergecast procedures, assuming that messages are transferred concurrently at each level, are the depth of the tree T, which would be at most $n - 1$.   □

4.7   Chapter Notes

We have seen various algorithms to build a spanning tree T of a graph G that can be used for basic communication operations such as broadcast and convergecast. Table 4.1 summarizes the performances of the algorithms investigated. It would be fair to say that *Flood*, which provides flooding only over an unstructured network, should be the last choice because it does not result in any tree structure that can be used for further communications. The algorithm *Term_ST* enhances

Table 4.1 Comparison of spanning tree algorithms

Algorithm	Description	Time Comp.	Msg. Comp.
Flood	Broadcasts a message to all nodes over an unstructured network	$\Theta(d)$	$O(m)$
Flood_ST	Broadcasts a message to all nodes and builds a spanning tree in process	$O(n)$	$O(m)$
Term_ST	Builds a spanning tree and nodes know when their part is over	$O(n)$	$O(m)$
Tarry_ST	Builds a spanning tree using a token	$\Theta(m)$	$\Theta(m)$
Bcast_ST	Broadcasts over a spanning tree and root knows all received	$\Theta(depth(T))$	$\Theta(n)$
Ccast_ST	Convergecasts over a spanning tree	$\Theta(depth(T))$	$\Theta(n)$

Flood_ST by providing a termination condition such that a node is aware that at least all nodes in its subtree have terminated with an additional d time units cost, which may be important in a network with a large d. A minor improvement to this algorithm would be the provision of an additional termination message from the root to inform all nodes that the algorithm is over. Broadcasting and convergecast-ing over T results in significant gains in both time and message complexities as shown.

For the nodes of a *MANET*, these algorithms could still be used, but they need to be activated periodically, say every H units. The magnitude of H would depend very much on the velocity of the mobile objects, and clearly, H would have to be a small time value when dealing with objects that are moving very fast. For a *WSN*, mobility is not a general concern, but fault tolerance and energy considerations are important. In order to provide fault tolerance so that nodes that have ceased functioning can be discarded, we would have to execute the spanning tree algorithm periodically again, this time H having a value many orders of magnitudes greater than the mobile counterpart. As most of the data processing in a *WSN* is performed in a special node called the *sink* with more sophisticated hardware and software facilities than the ordinary sensing nodes, broadcast and convergecast operations from and to the sink have important applications, and we will see alternative ways of performing these operations in *WSN*s in Part III.

In this chapter, we have not tried to optimize any property of the spanning tree, and forming a tree that covers all nodes of a graph was assumed sufficient. We will see in the next two chapters that some optimization such as a certain graph traversal method or the minimality of the total edge cost of the tree may be important for some applications. Spanning tree formation is a general topic treated in various books on distributed algorithms such as [5] and [2]. Broadcast, convergecast, and distributed spanning trees are listed as elementary algorithms in [1]. A formal presentation of the broadcast and convergecast algorithms is given in [3].

Fig. 4.7 Example Graph for
Exercise 1

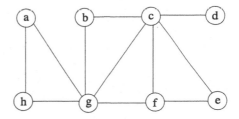

4.7.1 Exercises

1. Show possible message transactions and the formation of a spanning tree in
   Fig. 4.7 rooted at node *b* by Algorithm 4.2.
2. If Algorithm 4.2 is executed synchronously in the example graph of Fig. 4.7,
   what would be the characteristic of the resulting spanning tree and why?
3. Modify the state diagram of Algorithm 4.3 so that each node terminates in either
   ROOT, INTERM (a node having a child and a parent) or LEAF (a node without
   any children) states. Write a pseudocode that will provide the required operation.
4. Modify Algorithm 4.3 to provide synchronous operation.
5. Modify Algorithm 4.3 so that there are concurrent initiators, each being the root
   of the spanning tree to be formed. There will be *n* roots in this algorithm, and an
   array of states for each node may be used where each entry stores the state of a
   node in a tree of a specific root.
6. Write a pseudocode for the synchronous version of broadcast algorithm and work
   out its time and message complexities.

References

1. Gafni E (1986) Perspectives on distributed network protocols: a case for building blocks. In:
   Proc. IEEE MILCOM
2. Kshemkalyani K, Singhal M (2007) Distributed computing: principles, algorithms, and systems. Cambridge University Press, Cambridge
3. Segall A (1983) Distributed network protocols. IEEE Trans Inf Theory 29(1):23–35
4. Tary G (1895) Le problème des labyrinthes. Nouv Ann Math 14
5. Tel G (2000) Introduction to distributed algorithms, 2nd edn. Cambridge University Press,
   Cambridge

Chapter 5
Graph Traversals

Abstract This chapter introduces the basic distributed algorithms for breadth first search and depth first search in a graph. A spanning tree of the graph is formed after the execution of both algorithms.

5.1 Introduction

Traversal of a graph is performed by visiting all of its vertices in some predefined order. In an arbitrary graph G, it may be required to find the shortest distances of vertices from a source vertex r in terms of the number of links from r. A convenient way of achieving this is to proceed by layers, marking all neighbor vertices that are one hop away from r, then marking vertices that are one hop away from these neighbors, which are two hops away from r, and so on. The resulting tree rooted at r is called a *Breadth-First-Search* (*BFS*) tree of graph G. BFS may be used to find the connected components of a network; to find the shortest distances in terms of the number of hops between the nodes of a network or to test bipartiteness of a graph. A BFS tree of a graph G can be defined formally as follows.

Definition 5.1 (Breadth-First-Search Tree) A *breadth-first-search tree* T of a graph G is a spanning tree of G such that for every node of G, the tree path is a minimum-hop path to the root.

In order to find all of the vertices reachable from a source vertex r in a graph, *Depth-First-Search* (*DFS*) is used. Starting from a vertex r, DFS visits all possible vertices as far as it can reach, and when all vertices are visited, it returns to the parent node. A DFS tree of a graph G can be defined as follows.

Definition 5.2 (Depth-First-Search Tree) A *frond* edge is an edge that does not belong to a spanning tree. For a rooted spanning tree T of a graph G, let us denote by $S(u)$ all the nodes in the subtree of u, and $P(u)$ denote all the vertices that exist between u and the root. A *depth-first-search tree* of a graph G is a spanning tree T of G such that for every frond edge $\{u, v\}$, $v \in S(u) \vee v \in P(u)$ [5].

The DFS has many applications in distributed systems such as finding the strongly connected components of a directed network graph, which may be used to

K. Erciyes, *Distributed Graph Algorithms for Computer Networks*,
Computer Communications and Networks, DOI 10.1007/978-1-4471-5173-9_5,
© Springer-Verlag London 2013

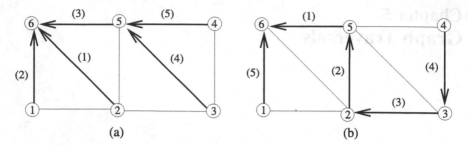

Fig. 5.1 (a) BFS tree. (b) DFS tree

detect cycles and hence prevent deadlocks in such systems. Also, a DFS algorithm may be used as the building block of various other distributed algorithms.

The operations of BFS and DFS are shown in Fig. 5.1, where each edge is labeled by the time step in which it is included in the tree. As shown in (b), $P(2) = \{5, 6\}$ and $S(2) = \{3, 4\}$ for node 2 on the DFS tree.

In this chapter, we describe two distributed BFS algorithms in Sect. 5.2 and three distributed DFS construction algorithms in Sect. 5.3.

5.2 Breadth-First-Search Algorithms

In this section, we show the implementation of two BFS algorithms. The first algorithm is synchronous working in rounds, and the second algorithm is asynchronous.

5.2.1 Synchronous BFS Construction

Our first distributed algorithm called *Synch_BFS* is the synchronous distributed version of Dijkstra's algorithm for the single-source shortest path problem. It has a single initiator, and in each synchronous round, a partial BFS tree is formed. The already formed branches of the tree are used to carry the synchronization messages and the leaves search for the new nodes to be added. The depth of the tree is incremented in each round until all nodes are covered.

As an introductory example, we will explicitly show all messages that are needed for the synchronous operation of this algorithm. Providing synchronization using special control messages as such eliminates the need for other synchronization methods like using a synchronizer. Also, full termination detection is provided so that all of the nodes know when the BFS tree is constructed and can be used. The following are the types of messages used in this algorithm:

- *round(k)*: Sent by the root at the beginning of round p and is transferred by all of the member nodes of the partial BFS tree to their children.

Algorithm 5.1 *Synch_BFS*

1: **int** *parent* $\leftarrow\perp$, $k \leftarrow 1$, *i*, *j*
2: **set of int** *childs, others, phased, finished* $\leftarrow \varnothing$
3: **message types** *round, probe, ack, reject, upcast*
4: **boolean** *leaf_flag* $\leftarrow$ *false*
5:
6: **if** *i* = *root* **then**
7:     **send** *probe(k)* to $\Gamma(i)$
8:     **while** *true* **do**
9:         **receive** *msg(j)*
10:         **case** *msg.type* **of**
11:             <u>*ack(k)* $\vee$ *upcast(k)*</u>:   *phased* $\leftarrow$ *phased* $\cup$ {*j*}
12:             <u>*finish*</u>:   *finished* $\leftarrow$ *finished* $\cup$ {*j*}
13:                                 **if** *finished* = *childs* **then**
14:                                     **send** *terminate* to *childs*
15:                                     **exit**   ▷ root terminates
16:                             **if** *phased* = *childs* **then**   ▷ start next round
17:                                 $k \leftarrow k+1$
18:                                 **send** *round(k)* to *childs*, *phased* $\leftarrow 0$
19:     **end while**
20: **else**
21:     **while** *true* **do**
22:         *round_over* $\leftarrow$ *false*
23:         **while** $\neg$*round_over* **do**
24:             **receive** *msg(j)*
25:             **case** *msg.type* **of**
26:                 <u>*round(k)*</u>:   **if** *state* = *interm* **then**   ▷ intermediate node
27:                                 **send** *round(k)* to *childs*
28:                             **else**   ▷ leaf node
29:                                 **send** *probe(k)* to $\Gamma(i) \setminus \{parent\}$
30:                             *round_recvd* $\leftarrow$ *true*
31:                 <u>*probe(k)*</u>:   **if** *parent* $=\perp$ **then**   ▷ non-member node
32:                                 *parent* $\leftarrow$ *j*; *state* $\leftarrow$ *new_leaf*
33:                                 **send** *ack(k)* to *j*
34:                             **else**   ▷ a member node
35:                                 **send** *reject(k)* to *j*
36:                 <u>*ack(k)*</u>:   *childs* $\leftarrow$ *childs* $\cup$ {*j*}
37:                             **if** *state* $\neq$ *interm* **then** *interm_flag* $\leftarrow$ *true*
38:                 <u>*reject(k)*</u>: *others* $\leftarrow$ *others* $\cup$ {*j*}
39:                             **if** *others* = $\Gamma(i) \setminus \{parent\}$ **then**   ▷ a leaf node finishes
40:                                 **send** *finish* to *parent*
41:                 <u>*upcast(k)*</u>: *phased* $\leftarrow$ *phased* $\cup$ {*j*}
42:                 <u>*finish*</u>:   *finished* $\leftarrow$ *finished* $\cup$ {*j*}
43:                             **if** (*finished* = *childs*) $\wedge$ (*state* = *interm*) **then**
44:                                 **send** *finish* to *parent*
45:                 <u>*terminate*</u>: **if** *childs* $\neq 0$ **then**
46:                                 **send** *terminate* to *childs*
47:                             **exit**   ▷ all nodes terminate

---

```
48:            if round_recvd then
49:                if ((state = leaf) ∧ ((childs ∪ others) = (Γ(i) \ {parent}))) ∨ ((state =
      interm) ∧ (childs = phased ∪ finished)) then                    ▷ check end of round
50:                    send upcast(p) to parent
51:                    if state = new_leaf then                              ▷ update states
52:                        state ← leaf
53:                    else if interm_flag then
54:                        state ← interm
55:                    end if
56:                    phased, others ← ∅; round_recvd ← false; round_over ← true
57:                end if
58:            end if
59:        end while
60:    end while
61: end if
```

- *probe(k)*: Sent by leaves of the partial BFS tree to unsearched neighbors.
- *ack(k)/reject(k)*: Sent by the searched node to accept/reject being a child of the sending leaf node.
- *upcast(k)*: Sent first by the leaf nodes of the partial BFS tree and then by the intermediate nodes to their parents, to signal the end of neighbor search and also the end of the round.
- *finish*: Sent by the leaf nodes to their parents to signal that their part is over as either they have no neighbors other than parent or they do not have any children.
- *terminate*: Broadcast by the root to all nodes to signal that the construction of the BFS tree is completed.

The root starts the algorithm by sending the first *round* message to its neighbors, which in turn respond by *ack* messages as shown in Algorithm 5.1. This message is transferred over the partial BFS tree to the leaf nodes, which in turn search nodes for the next level of the BFS tree by the *probe* messages. In order to detect the end of each round, each leaf node that has received *ack* or *reject* messages from all of its neighbors except the parent returns the *upcast* message to its parent, which in turn convergecasts *upcast* message to its parent. When all *upcast* messages from the neighbors of the *root* are received, root starts the next round by issuing the next *round* message.

The depth of the *BFS* tree formed would be $O(d)$, and the *root* could terminate all nodes after d rounds. However, this would require that the diameter of the network be known by the root beforehand. Closer inspection shows that when all the neighbors of a node i except its parent have responded by *reject* messages or when a leaf node does not have any neighbors other than its parent, part of i is over. Algorithm 5.1 uses this observation, and when the set of unrelated neighbors of a leaf node is equal to the set of all of its neighbors or if it does not have any neighbors other than the parent, it sends the *finish* message to its parent which is convergecast to *root*. When root receives the *finish* messages from all of its children, it further

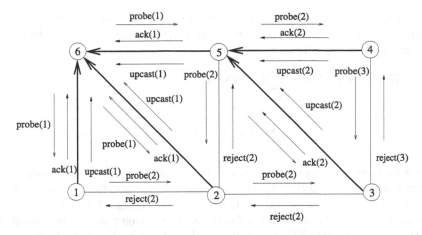

Fig. 5.2 *Synch_BFS* execution example

broadcasts a *terminate* message over the formed *BFS* tree to inform every node that the process is over.

5.2.1.1 Example

An example network in Fig. 5.2 shows six nodes numbered $1, \ldots, 6$, where node 6 initiates the algorithm by the *probe*(1) message. Each message is labeled by the round number it has transmitted. Each node that receives *probe* for the first time responds by the *ack* message and marks the sender as its parent. After two rounds, the *BFS* tree is formed as shown by bold arrows.

5.2.1.2 Analysis

Lemma 5.1 *Algorithm Synch_BFS correctly constructs a* BFS *tree.*

Proof We show this by induction. Assuming that, at step k, a *BFS* tree T_k rooted at r is formed, at step $k + 1$, only the leaves of the tree will be active in sending the probe message to their neighbors that are one hop away. Therefore any added nodes will be $k + 1$ hops away from r, forming T'_k.   □

Theorem 5.1 *The time complexity of Algorithm* 5.1, *Time(Synch_BFS), is* $O(d^2)$. *Its message complexity Msg(Synch_BFS) is* $O(nd + m)$ *where d is the diameter and m is the number of edges of graph G.*

Proof Broadcasting of the message *round*(k) to current T_k takes k units and convergecast of message *upcast* similarly is performed in k units, where k is the depth of T_k. There is also additional time for the *probe* and *ack/reject* messages at each step totalling $2k + 2$ time for each step. Therefore, assuming that the final *BFS* tree will be at most of depth d,

$$Time(Synch\_BFS) = \sum_k Time(Phase_k)$$

$$= 2 \sum_{k=1}^{d} (k + 1) = (d + 1)(d + 2)/2 = O(d^2). \qquad (5.1)$$

For the message complexity, we need to consider the synchronization messages and tree forming messages separately. In round p, if the tree formed has k nodes, there will be $k - 1$ *round* messages and $k - 1$ *upcast* messages for a total of $2k - 1$ messages providing a maximum number of synchronization messages in a round as $O(n)$. Since the formed BFS tree will have diameter d as the depth, the message complexity for the synchronization process is $O(nd)$. In each round, new edges of the BFS tree will be determined by the *probe* and *ack/nack* messages, and therefore, the total traversals for discovery of these edges will be at most $2m$. Summation of the synchronization and discovery messages yields:

$$Msg(Synch\_BFS) = \sum_p msg(Phase\ p) = O(nd) + O(m) = O(nd + m). \quad (5.2)$$

$\square$

5.2.2 Asynchronous BFS Construction

The second algorithm to build a BFS tree of a graph G is the distributed version of the Bellman–Ford algorithm called *Update_BFS*. We have a single initiator as before, which starts the algorithm by sending the *layer(l)* message that contains its distance to its neighbors as unity. Any node receiving a *layer(1)* message compares the layer value l contained in the message with its known distance to the root, and if the new value is smaller, the sender of the layer message is labeled as the new parent, and the distance is updated to l. Since the new distance to the root will affect all neighbors and other nodes, the *layer(l + 1)* message containing the new distance is sent to all neighbors except the new parent as shown in Algorithm 5.2. It can be seen that this process eventually builds a BFS tree starting from the root. The termination condition would be the traversing of the longest shortest path between any two nodes, which would be the diameter of the graph G.

5.2.2.1 Example

An example is shown in Fig. 5.3 with six nodes numbered $1, \ldots, 6$, where the *layer* message carries the distance, and the time frame it is delivered as *layer(distance,*

Algorithm 5.2 *Update_BFS*

1: **int** *parent* ← ∅, *my_layer* ← ∞, *count* = 1, *d* ← diameter of *G*
2: **set of int** *childs* ← ∅, *others* ← ∅
3: **message types** *layer*, *ack*, *reject*
4: **if** *i* = *root* **then**
5:       **send** *layer*(1) to *Γ*(*i*)
6: **end if**
7: **while** *count* ≤ *d* **do**
8:       **receive** *msg*(*j*)
9:       **case** *msg*(*j*).*type* **of**
10:             *layer*(*l*):   **if** *my_layer* > *l* **then**   ▷ update distance
11:                               *parent* ← *j*
12:                               *my_layer* ← *l*
13:                               **send** *ack*(*l*) to *j*   ▷ inform parent i am child
14:                               **send** *layer*(*l* + 1) to *Γ*(*i*)\{*j*}   ▷ update neighbors
15:                         **else**
16:                               **send** *reject*(*l*) to *j*   ▷ else reject sender
17:             *ack*(*l*):   *childs* ← *childs* ∪ {*j*}   ▷ include sender in children
18:             *reject*(*l*):   *others* ← *others* ∪ {*j*}   ▷ include sender in unrelated
19:       *count* ← *count* + 1
20: **end while**

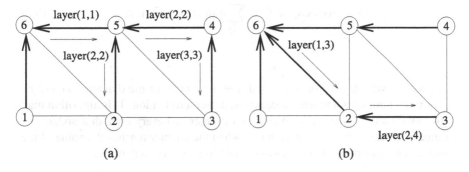

Fig. 5.3 *Update_BFS* execution example

time). Node 6 initiates the algorithm by sending the *layer*(1, 1) message to its one-hop neighbors. Each neighbor node, when it receives this message, compares the distance value in the message to its known distance and assigns its parent to the sender if the new distance is smaller. It can be seen in Fig. 5.3(a) that *layer* message reaches node 2 via node 5 before the direct connection between nodes 6 and 2, resulting in node 2 identifying node 5 as its parent. However, this situation is corrected in (b) when the layer message from node 6 reaches node 2 in the third time frame resulting in node 2 replacing its parent node 5 with node 6 correctly. Similarly, in time frame 4, node 3 replaces its parent node 4 with node 2 to correctly construct the BFS tree rooted at node 6.

5.2.3 Analysis

Lemma 5.2 *Algorithm Update_BFS correctly constructs a BFS tree.*

Proof After p steps, every vertex i that has a distance p to the root will have received the $layer(p-1)$ message from a neighbor, and it will set its variable my_dist to p. It will also choose parent j that has the distance $p-1$ to the root, and therefore a BFS tree will be formed after d steps. $\square$

Theorem 5.2 *The time complexity of Algorithm 5.2 is $O(d)$, and its message complexity is $O(nm)$.*

Proof Since the diameter d of G is the longest distance between any two nodes in G, all the *layer* messages will have reached all nodes in $O(d)$ time units. Hence, $Time(Update\_BFS) = O(d)$. Node v may have $n-1$ as the first value it ever assigns to its distance to the root. After that, it will change its distance value at most $n-2$ times as the longest path, and assuming each time it updates its distance to the root and sends *layer* messages to its neighbors in the worst case, it will have sent $n.deg(v)$ messages in total. Total number of messages therefore will be [3]

$$Msg(Asynch\_BFS) = \sum_{v=1}^{m} n.deg(v) = O(nm). \tag{5.3}$$

$\square$

A problem with this algorithm is that the magnitude of the diameter should be known by all nodes prior to the execution to detect termination. This algorithm may be improved so that each message carries a counter and every node that updates its distance increments the counter such that when the counter reaches the value of the diameter, the *layer* message is not transmitted to neighbors any more.

5.3 Depth-First-Search Algorithms

All the distributed algorithms considered in this section use a special message called *token*, which traverses the graph in DFS fashion always searching the graph as deep as possible by visiting unsearched neighbors of the current node, and if all these are searched, token is returned to the parent which is the first node that has sent the token. As the token provides a single point of activity at any time, the operation of the three algorithms is in fact sequential. We first show an algorithm that is an extension to Tarry's algorithm, the second one provides an improvement in time complexity to the first one, and the third algorithm includes identifiers of the nodes visited in the token.

Algorithm 5.3 *Classic_DFS*

```
 1: int parent ←⊥
 2: boolean visited[d_i] ← {false}
 3: message types token
 4:
 5: if i = root then                                    ▷ root starts the algorithm
 6:     parent ← i, choose j ∈ Γ(i)
 7:     visited[j] ← true, send token(i) to j
 8: end if
 9:
10: while true do
11:     receive token(j)
12:     if parent =⊥ then                                       ▷ token first time
13:         parent ← j
14:     end if
15:     if ∀q ∈ {Γ(i) \ {parent}}: visited[q]  then
16:         if i = root then                    ▷ if root and all searched, terminate
17:             exit
18:         elsesend token to parent
19:             visited[parent] ← true
20:             exit                               ▷ all nodes except root terminate
21:         end if
22:     else
23:         if j ≠ parent∧ ¬visited [j] then ▷ check to send token back from same channel
24:             q ← j
25:         else
26:             choose q ∈ {Γ(i) \ {parent}} : ¬visited[q]         ▷ send token to unsearched
                                                                             neighbor
27:         end if
28:         visited[q] ← true
29:         send token to q
30:     end if
31: end while
```

5.3.1 The Classical DFS Algorithm

Our first distributed DFS algorithm called *Classic_DFS* is obtained by modifying Tarry's algorithm that we have seen in Chap. 4 [4]. Tarry's algorithm had two rules: a process never forwards the token twice through the same channel, and a noninitiator forwards the token to its parent if it cannot forward it to any other neighbor with the first rule. This algorithm is formed by the addition of the following rule:

- R3: When a process receives the token, it sends it back through the same channel if this is allowed by Rules 1 and 2.

Algorithm 5.3 displays the operation of the Classical DFS Algorithm. Lines 23–24 implement R3 so that the token received via the frond edge is sent back to the sender.

Fig. 5.4 The classical DFS
algorithm execution example

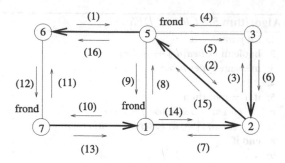

5.3.1.1 Example

Figure 5.4 shows the operation of *Classic_DFS* in a sample network with six nodes,
where the traversal of the token is displayed by the time frame it occurs along the
edges. Node 6 starts the algorithm, and after reaching node 2 in time 2, the token is
forwarded to node 3 in time 3, which forwards the token to its neighbor 5. Node 5,
however, applies Rule 3, finds edge {5, 3} that is a frond edge, and returns token
immediately back to 3. Similarly, token is returned to senders along the frond edges
{1, 5} and {6, 7}. It can also be seen that the construction of the whole DFS tree is
completed in 16 steps, which is twice the number of edges for this graph, as there
are two traversals for each edge.

5.3.1.2 Analysis

Theorem 5.3 *Algorithm 5.3 correctly constructs a DFS tree of an arbitrary
graph G in 2m times using 2m messages.*

Proof Using R3 ensures that the traversal of a frond edge is followed by the second
traversal in the opposite direction so that this edge is not included in the DFS tree.
Each edge is used to deliver a message twice, once in each direction for a total of
$2m$ times, resulting in $2m$ messages in total. Since there is a single activity at any
time, each message delivery action corresponds to a time unit resulting in $2m$ time
units.   □

A problem with this algorithm is that the frond edges that do not belong to
the DFS tree are also traversed by the token, resulting in a time complexity dependent on the number of edges m of the network graph. In general, m could be
as high as $O(n^2)$, and therefore, algorithms that have complexities dependent on
the number of nodes would be more desirable. Awerbuch's algorithm described
next achieves a linear time complexity dependent on the number of nodes of the
graph.

Algorithm 5.4 *Awerbuch_DFS*

```
 1: int parent ←⊥, i, j
 2: boolean visited[d_i] ← {false}
 3: message types token, vis, ack
 4: if i = root then                                          ▷ root starts the algorithm
 5:     parent ← i, send vis to Γ(i)
 6:     ∀j ∈ Γ(i), receive ack from j                        ▷ receive acks from all neighbors
 7:     choose q ∈ Γ(i)
 8:     send token to q
 9:     visited[q] ← true
10: end if
11: while true do
12:     receive msg(j)
13:     case msg(j).type of
14:         vis:     visited[j] ← true                       ▷ neighbor has token
15:                  send ack to j
16:         token:   if parent =⊥ then                       ▷ token first time
17:                     parent ← j
18:                     send vis to Γ(i)                      ▷ inform neighbors
19:                     ∀j ∈ Γ(i) \ {parent}, receive ack from j   ▷ receive acks
20:                  if ∀q ∈ {Γ(i) \ {parent}}: visited[q]
21:                     if i = root then exit
22:                     else send token to parent
23:                          visited[parent] ← true
24:                          exit                             ▷ root terminates
25:                  else choose q ∈ {Γ(i) \ {parent}}: ¬visited[q]
26:                     send token to q
27:                     visited[q] ← true
28: end while
```

5.3.2 Awerbuch's DFS Algorithm

Tarry's extended algorithm unnecessarily searches the nodes in frond edges. An algorithm that provides a token traversal in the DFS tree edges only will have time and message complexities in the order of n as the DFS tree will have $n - 1$ edges. Awerbuch provided an algorithm (*Awerbuch_DFS*) for this purpose, where a node receiving a token knows which of its neighbors has been visited previously by the token [1]. In this algorithm, there are also the notifying message that a node is visited (*vis*) and its response (*ack*) messages. When node i is visited for the first time, it sends a *vis* message to all of its neighbors except the parent. All neighbors of i send *ack* messages in return, and node i does not forward the *token* to any neighbor before receiving all of the *ack* messages as shown in Algorithm 5.4, where any neighbor of i that receives *vis* message marks its local data *visited*[i] as true to prevent sending of the token to that node in future.

5.3.2.1 Analysis

Theorem 5.4 *Awerbuch_DFS finds a DFS tree of an arbitrary graph G in $4n - 2$ time using $4m$ messages.*

Proof A tree edge is traversed four times, twice for token in each direction plus one for *vis* and one for *ack* messages. A nontree (frond) edge that is not included in the *DFS* will also be traversed four times as two *vis* messages and two *ack* messages in both directions. Therefore, the total number of messages traversed for all edges will be $4m$. The total time taken by the algorithm is twice the traversal of the tree formed as $2n - 2$, and also, for every node, there will be two units, one for sending *vis* and one for receiving *ack* messages. Therefore, the total time is $4n - 2$ time units. □

As an improvement, sending of a *vis* message to the neighbor which node will send the token can be omitted saving two messages (*vis* and *ack*) for each tree edge, resulting in the $2n - 2$ total message reduction.

5.3.3 Distributed DFS with Neighbor Knowledge

The DFS algorithm *Neigh_DFS* uses a token to traverse the nodes of the network in a sequential manner as in the previous algorithms. Since we need a way to know which nodes have been visited so that they are not visited again, a token may be used for this purpose. The token includes the visited node list which is appended by the node identifier of a node that is visited for the first time. The algorithm *Neigh_DFS* is depicted in Algorithm 5.5, where node i chooses a neighbor j to send the token only if it is not included in the list (*vislist*) of already visited nodes of the token.

5.3.3.1 Analysis

Theorem 5.5 *Neigh_DFS algorithm constructs a DFS tree in $2n - 2$ times using $2n - 2$ messages.*

Proof The final spanning tree will have $n - 1$ edges, and only the edges of this tree will have been traversed twice in each direction resulting in a total of $2n - 2$ token transfers among the nodes resulting in $2n - 2$ messages. Since there is a single activity at any time, each message delivery action corresponds to a time unit resulting in $2n - 2$ times. □

Figure 5.5 shows the operation of *Neigh_DFS* in a sample network, where n equals 8, and edges of the tree are labeled by the time frame token traverses them. The contents of the token when it is first received by a node and the final token as received by the root node 4 are also shown. The formed DFS tree takes 14 messages

---

Algorithm 5.5 *Neigh_DFS*

```
 1: int parent ←⊥
 2: set of int vislist ← ∅
 3: message types token
 4:
 5: if i = root then                                              ▷ root starts the algorithm
 6:     parent ← i, choose j ∈ Γ(i)
 7:     send token({i}) to j
 8: end if
 9:
10: while true do
11:     receive token(j, vislist)
12:     if parent =⊥ then                                         ▷ token received first time
13:         parent ← j
14:     end if
15:     if ∃j ∈ Γ(i) \ {vislist} then                            ▷ choose an unsearched node if any
16:         choose j ∈ Γ(i) \ {vislist}
17:         send token(vislist ∪ {i}) to j
18:     else if i = root then
19:         exit                                                  ▷ if all searched and root, terminate
20:     else                                ▷ if all searched and not root, return token to parent
21:         send token(vislist ∪ {i}) to parent
22:         exit                                                  ▷ all nodes except root terminate
23:     end if
24: end while
```

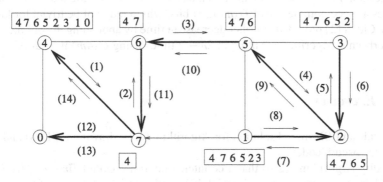

Fig. 5.5 *Neigh_DFS* execution example

$(2 \cdot 8 - 2)$ and also 14 time units as shown. The bit complexity of this algorithm is $O(n \log n)$, assuming that $\log n$ bits are used to represent a single node identity. For a large network where $n \gg 1$, this high bit complexity is disadvantageous as token has to be transferred $2n - 2$ times. For this example network, the bit complexity is $O(8 \cdot 3) = O(24)$.

Table 5.1 BFS and DFS
algorithm complexities

	Algorithm	Time Comp.	Msg. Comp.
BFS	Synch_BFS	$O(d^2)$	$O(nd + m)$
	Asynch_BFS	$O(d)$	$O(nm)$
DFS	Classical_DFS	$2m$	$2m$
	Awerbuch_DFS	$4n - 2$	$4m$
	Neigh_DFS	$2n - 2$	$2n - 2$

Fig. 5.6 Example graph for
Exercise 4

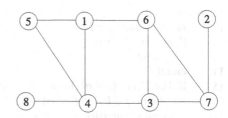

5.4 Chapter Notes

We have reviewed fundamental BFS and DFS algorithms. Comparison of BFS and
DFS Algorithms is shown in Table 5.1. For BFS, *Synch_BFS* has a better message
complexity, and *Asynch_BFS* has a better time complexity. A challenge, therefore,
is to design distributed BFS algorithms that optimize both time and message complexities. For DFS, the *Neigh_DFS* performs better than the other two algorithms
as in general, $m \gg n$; however, the token has to hold the identifiers of nodes, which
requires a space complexity of $O(n \log n)$ bits, which may not be trivial for large n
values. Cidon [2] provided a further decrease in time by abolishing the *ack* messages
in Awerbuch's Algorithm resulting in $2n - 2$ times using $O(4m)$ messages.

5.4.1 Exercises

1. Provide an FSM-based solution for Algorithm 5.2 by drawing the FSM and writing the pseudocode.
2. Modify Algorithm 5.2 so that a counter, sometimes called *Time To Live* (*TTL*)
   field, in each message is used, which is initialized to the diameter of the network
   by the root. Each node receiving the *layer* message now decrements TTL value,
   and if this reaches zero, the *layer* message is not forwarded to any neighbors.
   Show the operation on the sample graph of Fig. 5.2.
3. Provide the pseudocode for the synchronous version of *Update_BFS* algorithm
   and work out its time and message complexities.
4. Show the execution of the sequential BFS and DFS algorithms in the sample
   graph of Fig. 5.6 starting from node 1. The DFS algorithm always selects the
   highest identifier node from the unsearched nodes.

Fig. 5.7 Example graph for Exercise 4

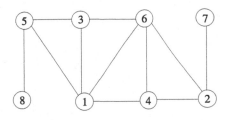

Fig. 5.8 Example graph for Exercise 5

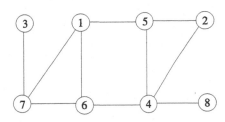

5. Show the execution of *Asynch_BFS* Algorithm in the example graph of Fig. 5.7 starting from node 3.
6. Describe and show the messages in a possible execution of *Neigh_DFS* Algorithm in the example graph of Fig. 5.8 starting from node 6 and assuming highest identifier is always selected to transfer the token. Show also the contents of the token in each iteration.
7. Describe how *ack* messages *in Awerbuch_DFS* Algorithm may be prevented (Cidon's Algorithm). Also, show that this algorithm achieves $2n - 2$ times using at most $4m$ messages.

References

1. Awerbuch A (1985) A new distributed depth first search algorithm. Inf Process Lett 20:147–150
2. Cidon I (1988) Yet another distributed depth-first-search algorithm. Inf Process Lett 26:301–305
3. Peleg D (2000) Distributed computing: a locality-sensitive approach, Chap. 5. SIAM, Philadelphia. ISBN 0-89871-464-8
4. Tary G (1895) Le problème des labyrinthes. Nouv Ann Math 14
5. Tel G (2000) Introduction to distributed algorithms, 2nd edn. Cambridge University Press, Cambridge

Chapter 6
Minimum Spanning Trees

Abstract A minimum spanning tree of a weighted graph is its spanning tree T with a minimum total cost of edges in T of all possible spanning trees. Minimum spanning trees have many applications in computer networks. In this chapter, we investigate synchronous and asynchronous distributed algorithms to construct minimum spanning trees.

6.1 Introduction

A *minimum spanning tree* (*MST*) of a weighted graph $G(V, E, w)$, where $w : E \rightarrow \mathbb{R}$, is a spanning tree such that the sum of the weights of edges of this tree is minimum among all possible spanning trees of the graph G. If G is unconnected, a *minimum spanning forest* of G is the union of all its such trees. Formally:

Definition 6.1 Given a weighted graph $G(V, E, w)$, $MST(G) = (V, E', w')$ with $E' \subset E$ that minimizes $\sum_e w'_e \in E'$.

MSTs can be used in constructing networks between nodes using the least amount of communication wire, in smaller electronic circuits, and for clustering. MST of a graph may be also used for more complicated algorithms and protocols. The MST algorithm also solves the leader election problem in general graphs, where the leader is simply the last root. In this chapter, we first describe two sequential algorithms and then analyze distributed implementations based on these algorithms.

6.2 Sequential MST Algorithms

Various sequential algorithms exist to build an MST of G. *Kruskal's Algorithm* (*Kruskal_MST*) starts with an empty MST and selects edges to include in MST from E such that the lowest weight edge that does not produce a cycle with the already selected edges is chosen. For the example network of Fig. 6.1, the selected edge sequence would be with weights 1, 2, 3, 4, and 7 in ascending order to give the MST excluding edges 5 and 6 as they produce cycles with the existing tree structure. *Kruskal_MST* has a time complexity of $O(m \log m)$ as it has to sort the edges

K. Erciyes, *Distributed Graph Algorithms for Computer Networks*,
Computer Communications and Networks, DOI 10.1007/978-1-4471-5173-9_6,
© Springer-Verlag London 2013

Fig. 6.1 Kruskal Algorithm
example. MST is shown by
bold lines

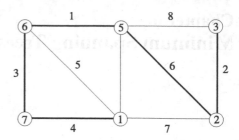

first. A quick observation is that *Kruskal_MST* is more convenient for distributed processing as it chooses edges independently.

The second sequential algorithm is due to Prim (*Prim_MST*) and executes in a greedy manner by always choosing the lightest weight edge that is outgoing from the current tree. We need to define few concepts to illustrate *Prim_MST* in detail.

Definition 6.2 A fragment F of an MST is a subtree of the MST. An outgoing edge $\{u, v\}$ of F is an edge such that $u \in F$ and $v \notin F$. The minimum-weight outgoing edge of a fragment of the MST is such an outgoing edge.

There are two important properties of MSTs as follows:

Lemma 6.1 (Property 1: blue rule) *Given a fragment of an MST, let e be a minimum-weight outgoing edge of the fragment. Then joining e and its adjacent nonfragment node to the fragment yields another fragment of an MST.*

Proof Assume that e is MWOE from the MST fragment $F_p(V_p, E_p)$ and T_m is the MST. If $e \notin T_m$, then there exists $e' \in T_m$ that is included in the cut $(V, V - V_p)$. Substituting e for e' in T_m results in a lighter spanning tree T' as $w(e) < w(e')$, which means that T_m was not minimum and thus that the original assumption $e \notin T_m$ was wrong, resulting in a contradiction.   □

Lemma 6.2 (Property 2) *If all edges of a connected graph have unique weights, then the MST is unique.*

Algorithm by Prim (*Prim_MST*) iteratively chooses the MWOE from the current MST fragment. Lemma 6.1 assures correct construction of the MST. Applying *Prim_MST* to the graph of Fig. 6.1 would give the same MST as before as there is one MST for a graph with unique weights. However, the order of edges included in the MST would be different as 1, 3, 4, 7, and 2, starting from node 6, since this algorithm chooses the MWOE from the existing vertices of the fragment of MST. *Prim_MST* using adjacency matrix requires $O(n^2)$ time which can be reduced to $O(m \log n)$ steps using binary heaps.

In this chapter, we will describe a synchronous MST algorithm based on *Prim_MST*, another synchronous algorithm based on *Kruskal_MST*, and then show a detailed implementation of an asynchronous algorithm.

6.3 Synchronous Distributed Prim Algorithm

The first synchronous algorithm is the distributed version of *Prim_MST* executed synchronously, called *DistPrim_MST*. There is a single initiator that completes the structuring of the tree in a number of rounds. In each round, the root gathers all MWOEs from the leaves of the partial tree T', finds the minimum $\{u, v\}$ of these MWOEs, and broadcasts $\{u, v\}$ to T' in the next round so that it can be added to T'. In this implementation, we will not use an FSM explicitly but will assign states to every node in the graph. Each node other than the root can be in intermediate (*interm*) node, *leaf $_p$* of the partial tree T', a node in $\{V \setminus T'\}$ (*other*), or a final leaf (*leaf $_f$*) state. At the reception of a *round* message, each node acts differently based on its state as follows:

1. An *interm* node simply forwards the *round*(k, x, y) message with the assigned MWOE information to its children during a *downcast* operation and converge-casts the lightest MWOE among the MWOEs it receives from its children to its parent during an *upcast* operation.
2. A *leaf $_p$* node that is connected to the assigned MWOE from a previous round sends the *round*(k, x, y) message to the node y connected to the other end of the MWOE and updates its data structures.
3. A *leaf $_p$* node that is not connected to the assigned MWOE from a previous round sends the *probe* messages to its *unexplored* neighbors, chooses the lightest edge from the neighbors that have responded with an *ack* message, and sends its MWOE to its parent. If all of its unexplored neighbors have responded with the *reject* message, it enters the *leaf $_f$* state and sends an *upcast* message with a *null* value indicating that it has terminated.
4. An *other* node may receive a *round*(k, x, y) message destined to it as it is at the other end of MWOE, in which case it enters the *leaf $_p$* state and updates its data structures. An *other* node may receive a *probe* message from a *leaf $_p$* node, in which case it sends an *ack* to the sender if it is the first *probe* message it receives; otherwise, it replies with a *reject*. Some *other* nodes may not receive any messages in some rounds if they are far from the root.

The messages used in the algorithm are the above-mentioned *round*, *assign*, *probe*, *ack*, *reject*, and *upcast* messages. The important data structures in this algorithm are as follows:

- *neigh_edges*: Set of tuples as $\langle u, v, w_{uv} \rangle$ to hold the neighbor unexplored edges.
- *down_mwoes*: Set of tuples as $\langle u, v, w_{uv} \rangle$ for the MWOEs received from the children.
- *unexplored*: Set of the identities of neighbor nodes that are connected to incident unexplored edges.

The termination condition for the algorithm is that when all nodes including the root cannot find any *MWOEs* that is disclosed to the root by sending a *null* value for the MWOE nodes as shown in Algorithm 6.1.

Figure 6.2 shows an example network where rounds are shown in parentheses and node 4 is the root node, which initiates the algorithm by sending the first *round*

Algorithm 6.1 *DistPrim_MST*

```
 1: int parent; state ← other
 2: set of int childs, responded, converged ← ∅; unexplored ← Γ(i)
 3: message types round, probe, ack, reject, upcast
 4: boolean terminated, round_recvd, round_over ← false
 5: if i = root then
 6:     find MWOE incident to j ∈ Γ(i)
 7:     send round(1, root, j) to j
 8: end if
 9: while ¬terminated do
10:     round_over ← false
11:     while ¬round_over do
12:         receive msg(j)
13:         case msg.type of
14:             round(k, x, y):    if state = leaf_p then
15:                                    if (i = x) then                           ▷ MWOE parent
16:                                        childs ← childs ∪ {y}
17:                                        lost_neighs ← lost_neighs ∪ {j}
18:                                        send round(k, x, y) to y
19:                                        if unexplored ≠ 0 then
20:                                            send probe to unexplored
21:                                    else if state = interm then
22:                                        send round(k, x, y) to childs
23:                                    else if i = y then                         ▷ MWOE child
24:                                        parent ← x, state ← leaf_p
25:                                        lost_neighs ← lost_neighs ∪ {j}
26:                                    round_recvd ← true
27:             probe(k):          if state = other ∧ ¬searched then
28:                                    send ack(k) to j
29:                                    parent ← j; searched ← true
30:                                else send reject(k) to j
31:             ack(k):            responded ← responded ∪ {j}
32:                                neigh_edges ← neigh_edges ∪ {i, j, w_ij}
33:             reject(k):         lost_neighs ← lost_neighs ∪ {j}
34:                                responded ← responded ∪ {j}
35:                                if lost_neighs = Γ(i) \ {j} then
36:                                    state ← leaf_p; send upcast(k, 0, 0) to parent
37:             upcast(k, u, v):   converged ← converged ∪ {j}
38:                                down_mwoes ← down_mwoes ∪ {u, v, w_uv}
39:
40:         if  round_recvd ∧ ((state = root/interm ∧ (converged = childs)) ∨ ((state =
            leaf_p) ∧ (responded = unexplored))) then
41:             if ∃(u, v, w_uv) ∈ {(neigh_edges ∪ down_mwoes)|u ≠ null} then
42:                 (p, q, w_pq ← min{w|(u, v, w_uv) ∈ neigh_edges ∪ down_mwoes}
43:             if i ≠ root then
44:                     send upcast(p, q, w_pq) to parent
45:             else k ← k + 1
46:                     send round(k, p, q) to childs
47:         end if
```

```
48:            else terminated ← true
49:                if i ≠ root then
50:                    send upcast(k, null, null) to parent
51:                end if
52:            end if
53:            unexplored ← unexplored \ lost_neighs
54:            responded, converged, lost_neighs, neigh_edges, down_mwoes ← ∅
55:            round_recvd ← false; round_over ← true
56:        end if
57:    end while
58: end while
```

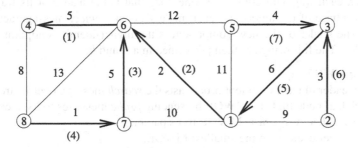

Fig. 6.2 *DistPrim_MST* execution example

message to its MWOE neighbor 6, to be included in the tree. Node 6 in return upcasts its MWOE of weight 2, node 4 has MWOE node of weight 8, and edge {6, 1} is chosen as the next edge to be included in the MST. The process goes on, and after seven rounds, there is no MWOE left, and the algorithm terminates.

6.3.1 Analysis

Theorem 6.1 *DistPrim_MST correctly constructs an MST of the graph G in $O(n^2)$ time, and its message complexity is also $O(n^2)$.*

Proof The correctness is ensured by Lemma 6.1. The MST formed will have $O(n)$ as its depth since there will be $n - 1$ nodes in the longest path. The time required for a round and also the number of rounds are therefore both $O(n)$. The total time in this case is $O(n^2)$. Each round requires $O(n)$ messages as the MST will have $O(n)$ edges. Summing over all rounds gives $O(n^2)$ for the total number of messages. ☐

6.4 Synchronous GHS Algorithm

Gallager, Humbelt, and Spira proposed a synchronous algorithm (*SGHS_MST*) with concurrent initiators to find the MST of a graph. The general idea of this algorithm is to have the initiators expand their own fragments synchronously in phases by finding their MWOEs of the fragments and joining fragments in parallel. This algorithm requires synchronization of each phase, which may be provided by a synchronizer, or a spanning tree of the whole network may be constructed beforehand as in the previous algorithm of *Prim_Synch* at the cost of additional messages. Additionally, differently from the synchronous algorithms we have seen before, this algorithm also requires synchronization of all fragments within a round so that nodes in each fragment should be executing the same round concurrently.

Initially, each fragment consists of one node that is the leader of its fragment. In the first phase of the algorithm, each node sends a *connect* message across its MWOE. The leader of the new component is the higher-identity endpoint of this unique edge. The following are then performed in a round k:

1. *Finding MWOE*:
   a. The leader of each fragment broadcasts the *search* message over its tree T.
   b. Each leaf node finds its MWOE by sending *probe* messages to its unexplored edges and convergecasts MWOE to its parent.
   c. The leader decides on the smallest MWOE.
2. *Combining Fragments*: Assuming that fragments have T_1 and T_2:
   a. The leader sends a *connect* message to the leaf node with MWOE
   b. One of the endpoints of MWOE with the higher identity is chosen as the new leader.
   c. The T_c is formed by modifying the T_1 and T_2 links so that they point to the new leader.
   d. The new leader broadcasts the *new_leader* message to all nodes in the combined fragment so that they can use this identifier subsequently.

The implementation of this algorithm will be similar to *DistPrim_MST*. Figure 6.3 shows an example network where the initial fragments F_8, F_7, F_5, and F_2 have nodes 8, 7, 5, and 2 as the roots of their trees as shown in (a). F_8 and F_7 merge through edge $\{4, 6\}$, electing a new leader as node 6; F_5 and F_2 merge through edge $\{1, 3\}$, electing a new leader as node 3 as these nodes have the higher identifiers of the merged edges as shown in (b) to form fragments F_6 and F_3. The final MST is formed by merging these two fragments under the final leader node 7 as shown in (c). In this example, the weights are chosen such that there is at least one neighbor v of node u such that MWOE(u) = MWOE(v) initially. It may be possible that the MWOE of a node may not be the MWOE of none of its neighbors, in which case it may have to wait for another round before joining a fragment.

Figure 6.4 displays such an example where nodes 4 and 5 have MWOEs ($\{4, 6\}$ and $\{5, 3\}$) that are not MWOEs of any of their neighbors. When a node has edges that are MWOEs of more than one neighbor, the highest identifier node is accepted in this implementation.

Fig. 6.3 *Synch_GHS example execution 1*

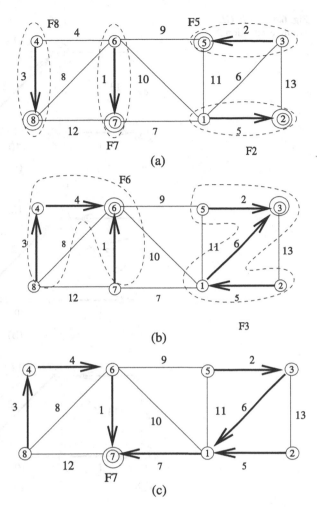

(a)

(b)

(c)

6.4.1 Analysis

Theorem 6.2 *The time complexity of SGHS_MST is $O(n \log n)$, and its message complexity is $O(m \log n)$.*

Proof In each round of the algorithm, there is a broadcast and convergecast for MWOE. As the size of a fragment is at most n, the time for broadcast and converge-cast is $O(n)$, and message complexities for these two operations are also $O(n)$ since an n-node tree would have at most $n - 1$ edges and the fragment merging has similar complexities. However, in the final phase of the fragment merging step (Step 2.d), the new fragment identifier is updated by each node resulting in $\Theta(m)$ messages.

Fig. 6.4 *Synch_GHS* example execution 2

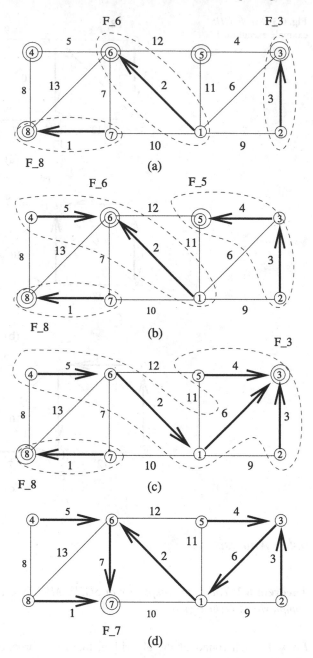

As there will be $O(\log n)$ rounds, the total time complexity is $O(n \log n)$, and the message complexity is $O(m \log n)$ considering the last step of merging, which is the dominant message transfer cost as $m \gg n$ in general [7]. □

6.5 Asynchronous GHS Algorithm

Using the same idea of independently growing fragments and joining them, Gallager, Humbelt, and Spira proposed an asynchronous version of the *SGHS_MST* algorithm, which we call *AGHS_MST*. There are several issues to be addressed in the asynchronous version of the algorithm as follows. First, nodes do not have complete information about their MWOEs. In *SGHS_MST*, nodes are at the same level when they are testing their edges, but in this algorithm neighbors may be out of phase. Also, in *SGHS_MST*, level k fragments have 2^k nodes, and level $k+1$ fragments are constructed from at least two level k fragments, whereas in *AGHS_MST*, components at different levels can be combined. There is a need for a more sophisticated protocol.

The general idea of the algorithm is that each node i is a fragment initially containing itself and the fragments are enlarged using the blue rule as before. Each fragment F_i finds its MWOE asynchronously, and when MWOE is determined, F_i is combined with F_j at the other end of MWOE using a protocol, and the algorithm terminates when there is only one fragment. The following are the messages used in *AGHS_MST*:

- *initiate*: Broadcast from the leader to find MWOE.
- *report*: Convergecasts MWOE responses back to the leader.
- *test*: Asks whether an edge is outgoing from the component.
- *connect*: Sent across the MWOE to connect components.
- *changeroot*: Sent from the leader to the endpoint of MWOE.

A *merge* occurs when the *connect* message has been sent both ways on the edge, where two nodes must have the same level, and an *absorb* occurs when the *connect* message has been sent on the edge from a lower-level to a higher-level node.

6.5.1 States of Nodes and Links

Each node can be in three states as *Sleeping*, *Find*, and *Found*, where *Sleeping* is the initial state of all nodes. A node will be in state *Find* when searching MWOE and in state *Found* at all other times. A state of an each edge $e = \{u, v\}$ can be one of the following:

- *Basic*: It has not yet been decided whether the edge is part of the MST or not.
- *Rejected*: The edge is not part of the MST.
- *Branch*: The edge is part of the MST.

6.5.2 Searching MWOE

In this section, we analyze the searching of MWOE by fragments. The first step is realized by the *Test-Accept-Reject Protocol*, which determines MWOE, and then combining of the fragments can be achieved as described below.

6.5.2.1 Test-Accept-Reject Protocol

Every node that receives the *initiate* message starts searching its MWOE by sending out the *test*(F_i, L_i) message on its minimum-weight basic edge, where F_i and L_i are the fragment identifier and the level of the sending node i. A node with fragment identifier F_j and level L_j that receives the *test* message does the following:

- if $F_i = F_j$, the *reject* message is sent to i.
- if $(F_i \neq F_j) \wedge (L_i \leq L_j)$, the *accept* message is sent to i.
- if $(F_i \neq F_j) \wedge (L_i > L_j)$, no reply is sent until this condition changes.

Whenever a node receives the *reject* message from a neighbor node as a result of the *test* message, it marks the edge as *rejected* and sends the *test* message to the next minimum-weight basic edge. After receiving responses from all the neighbors that *initiate* was sent, the node sends the *report* message to the node it received the *initiate* message. All of the *report* messages are convergecast to the *core* node that is the leader of the fragment. After core nodes find MWOEs, the core node that is closer to the MWOE sends the *changeroot* message over the path from the core to the MWOE connected node, which in turn sends the *connect*(L_i) message to the node in F_j incident to MWOE.

6.5.2.2 Connecting Fragments

Each fragment F_i has a level L_i, which is initially set to 0. When a smaller level fragment combines into a larger one, it changes its label and level to those of the larger fragment. If two fragments at level L have the same MWOE, each sends the *connect* message over this MWOE, and the edge becomes the *core* of the new fragment of level $L + 1$ and the *initiate* messages are sent to find the new MWOE. If the *connect* message is exchanged between nodes of different levels, say $L_1 < L_2$, of fragments F_1 and F_2, the node incident to MWOE at L_2 immediately sends the *initiate* message that is broadcast to all nodes in L_1, in which case L_2 becomes the level of all nodes in F_2. In this case, F_1 is said to be *absorbed* by F_2. Formally, combining method for fragments is as follows:

1. **Rule A**: If MWOE$_i$ of F_i is connected to F_j, where $L_i < L_j$, $F_i \leftarrow F_j$, and $L_i \leftarrow L_j$; the new values of F_i and L_i are broadcast to all nodes of F_i.
2. **Rule B**: If MWOE$_i$ of F_i is connected to F_j, where $L_i = L_j$, F_i and $F_j \leftarrow F_w$ and L_i and $L_j \leftarrow L_i + 1$; $L_i \leftarrow L_j$; the new values of F_i and L_i are broadcast to all nodes of F_i and F_j.
3. **Rule C**: For all other cases, F_i must wait until Rule A or Rule B is applicable.

6.5.3 The Algorithm

Algorithm 6.2 shows the operation of *AGHS_MST* as in [8] as regards to the above description of its operation. The procedure *do_test* is invoked when a node receives an initiate message, *do_report* is invoked when a node receives a *report* message, and *do_changeroot* is used by a node to change its root.

6.5.3.1 Analysis

Theorem 6.3 *AGHS_MST computes an MST using $O(m \log n)$ messages in $O(n \log n)$ time.*

Proof Each step of *AGHS_MST* requires a broadcast and convergecast of messages, and then a step of combining fragments is performed. These steps require $O(n)$ time, as the depth of a fragment is at most n, and $O(n)$ messages, as the number of edges in a fragment is also at most n. The combining of the fragments requires similar time and message complexities as the messages traverse in the fragment. However, in the last step, there are at most $\Theta(m)$ messages for updating the new core identifier. Since the number of phases is $O(\log n)$, the total number of time steps considering $O(n)$ steps per phase is $O(n \log n)$. The message complexity therefore, counting both the initial and combining steps, is $O(m + n \log n)$.

In detail, there will be $4|E|$ test-reject messages (one pair for each side of every edge), n initiate messages per level, n report messages per level, $2n$ (*test-accept*) messages per level (one pair for each node), n (*change-root/connect*) messages per level (core to MWOE path), addition of which yields $4m + 5n \log n$ messages, and therefore, $Msg(AGHS\_MST) = O(m + n \log n)$.   □

6.6   Chapter Notes

The distributed MST problem is a fundamental problem in distributed computing. *AGHS_MST* algorithm, which has $O(n \log n)$ time complexity and $O(m + n \log n)$ message complexity, was a fundamental asynchronous distributed MST algorithm, which has inspired further research. Chin and Ting [2] improved the time complexity of GHS algorithm to $O(n \log \log n)$, Gafni [3] provided a further improvement to $O(n \log^* n)$, and then Awerbuch [1] provided a running time of $O(n)$, which is optimal.

Considering the diameter d of the graph, Garay, Kutten, and Peleg [4] provided a distributed MST algorithm with running time $O(d + n^{0.61})$. Kutten and Peleg [6] further improved the bound to $O(d + \sqrt{n} \log^* n)$. Maleq and Pandurangan [5] viewed the problem from the point of providing an approximate MST algorithm rather than an exact one, and they proposed an $O(\log n)$-approximate distributed MST algorithm with running time $O(d + l)$, where d is the diameter, and l is the *local shortest path diameter* of the graph, which depends on the topology and the edge weights of the graph.

Algorithm 6.2 *AGHS_MST*

1: **state**: (*sleep, find, found*)
2: **int** *statech*[|$\Gamma(i)$|]: *basic, branch, reject*
3: **int** *level, rec*
4: **neighbors** *testch, bestch, parent*
5: **real** *name, bestwt*
6: **message types** *initiate, connect, test, ack, reject, report*
      {Initialize}
7: find $MWOE(\{i, k\})$
8: *state* ← *found, level* ← 0; *statech*[k] ← *branch, rec* ← 0
9: **send** *connect* to k
10: **while** *true* **do**
11:     **receive** *msg*(j)
12:     **case** *msg.type* **of**
13:         *connect*(L):   **if** L < *level* **then**
14:                               *statech*[j] ← *branch*
15:                              **send** *initiate*(*level, name, state*) to j
16:                     **else if** *statech*[j] = *basic* **then** process message later
17:                          **else send** *initiate*(*level* + 1, $\omega(ij)$, *find*) to j
18:         *initiate*(L, F, S): *level* ← L; *name* ← F; *state* ← S; *parent* ← j;
19:                          *bestch* ←⊥; *bestwt* ← ∞;
20:                          **for all** $x \in \Gamma(i)$: *statech*[x] = *branch* $\wedge$ $x \neq j$ **do**
21:                               **send** *initiate*(L, F, S) to x
22:                          **if** *state* = *find* **then** *rec* ← 0; *do_test*
23:         *test*(L, F):   **if** (L > *level*) **then** process message later
24:                          **else if** F = *name* **then**
25:                                 **if** *statech*[j] = *basic* **then** *statech*[j] ← *reject*
26:                               **if** $j \neq$ *testch* **then send** *reject* to j
27:                               **else** *do_test*
28:                         **send** *accept* to j
29:         *accept*():   *testch* ←⊥
30:                          **if** $\omega(ij)$ = *bestwt* **then**
31:                             *bestwt* ← $\omega(ij)$; *bestch* ← j
32:                        *do_report*
33:         *reject*():   **if** (*statech*[j] = *basic*) **then** *statech*[j] = *reject*
34:                        *do_test*
35:         *report*(ω):   **if** ($j \neq$ *parent*) **then**
36:                          **if** ω < *bestwt* **then**
37:                                 *bestwt* ← ω; *bestch* ← j
38:                        *rec* ← *rec* + 1; *do_report*
39:                     **else**
40:                        **if** *state* = *find* **then** process message later
41:                        **else if** ω > *bestwt* **then** *do_changeroot*
42:                          **else if** ω = *bestwt* = ∞ **then terminate**
43:         *changeroot*(): *do_changeroot*
44: **end while**

```
45: procedure do_test
46:     if ∃ j ∈ Γ(i) : statech[ j] = basic then
47:         testch ← j ∧ statech[ j] = basic ∧ ω(ij) minimal
48:         send  test(level, name) to testch
49:     else testch ←⊥; do_report
50:     end if
51: end procedure
52:
53: procedure do_report
54:     if rec = #{ j : statech[ j] = branch ∧ j ≠ parent} ∧ testch =⊥ then
55:         state ← found; send  report(bestwt) to parent
56:     end if
57: end procedure
58:
59: procedure do_changeroot
60:     if  statech[bestch] = branch then
61:         send  changeroot to bestch
62:     elsesend  connect(level) to bestch; statech[bestch] ← branch
63:     end if
64: end procedure
```

Fig. 6.5 Example graph for
Exercise 1

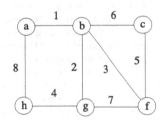

Fig. 6.6 Example graph for
Exercise 2

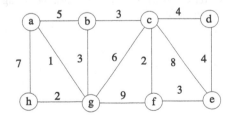

6.6.1 Exercises

1. Find MST in the graph of Fig. 6.5 using *Kruskal_MST* and *Dijkstra_MST* algorithms.
2. Show the execution of *DistPrim_MST* algorithm in the graph of Fig. 6.6 assuming that node *a* is the root.
3. Show a possible execution of *SGHS_MST* in the graph of Fig. 6.7.

Fig. 6.7 Example graph for
Exercise 3

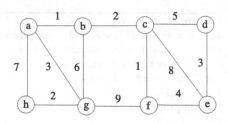

4. Under which circumstances would an approximate distributed MST algorithm
   be preferred over an exact distributed algorithm?

References

1. Awerbuch B (1987) Optimal distributed algorithms for minimum weight spanning tree, counting, leader election, and related problems. In: Proc 19th ACM symp on theory of computing,
   pp 230–240
2. Chin F, Ting H (1985) An almost linear time and $O(n \log n + e)$ messages distributed algorithm
   for minimum-weight spanning trees. In: Proc 26th IEEE symp foundations of computer science,
   pp 257–266
3. Gafni E (1985) Improvements in the time complexity of two message-optimal election algorithms. In: Proc of the 4th symp on principles of distributed computing, pp 175–185
4. Garay J, Kutten S, Peleg D (1998) A sublinear time distributed algorithm for minimum-weight
   spanning trees. SIAM J Comput 27:302–316
5. Khan M, Pandurangan G (2008) A fast distributed approximation algorithm for minimum spanning trees. Distrib Comput 6(20):391–402
6. Kutten S, Peleg D (1998) Fast distributed construction of k dominating sets and applications.
   J Algorithms 28:40–66
7. Peleg D (2000) Distributed computing: a locality-sensitive approach. SIAM, Philadelphia.
   ISBN 0-89871-464-8
8. Tel G (2000) Introduction to distributed algorithms, 2nd edn. Cambridge University Press,
   Cambridge

Chapter 7
Routing

Abstract Routing in a computer network is the process of communicating messages from source nodes to destination nodes along selected paths with the lowest possible costs. This chapter introduces few sample distributed routing algorithms based on sequential routing algorithms.

7.1 Introduction

For the routing process, we will assume that the edges of G have nonnegative weights and that these represent costs of sending the messages and delays incurred. The network graph in this case is represented by weighted communication links as $G(V, E, w)$, where $w : E \rightarrow \mathbb{R}$. Since nodes are not connected to every other node of the weighted graph, messages must be forwarded between the intermediate nodes from source to the destination. The cost of sending a message from a source to a destination is the sum of the weights of the edges of the path between them. There is at least one shortest path between every pair of nodes, and the purpose of a routing algorithm is to determine this shortest path. Desirable properties of a routing algorithm are as follows:

- *Correctness*: Every message should be delivered correctly to its destination.
- *Complexity*: The algorithm must have low time, message, and space complexities.
- *Robustness*: The algorithm should update routing tables when topology changes.
- *Shortest Paths*: Messages should be transferred along the minimum-cost paths from the source to the destination.

In this chapter, we will first review three classical sequential routing algorithms due to Dijkstra, Bellman and Ford, and Floyd and Warshall in Sect. 7.2 as these form the basis of the distributed routing algorithms. Then we describe distributed implementations of Bellman–Ford and Floyd–Warshall algorithms in Sects. 7.3, 7.4, 7.5, and 7.6. We conclude by the descriptions of two fundamental routing protocols.

7.2 Sequential Routing Algorithms

In this section, two Single-Source Shortest-Path (*SSSP*) algorithms, which execute sequentially to find the shortest path from a single source to all nodes of the network,

K. Erciyes, *Distributed Graph Algorithms for Computer Networks*,
Computer Communications and Networks, DOI 10.1007/978-1-4471-5173-9_7,
© Springer-Verlag London 2013

Algorithm 7.1 *Dijkstra_SSSP*

1: $d_s \leftarrow 0$ ▷ initialize distances
2: **for all** $i \neq s$ **do**
3:     $d_i \leftarrow \infty$
4: **end for**
5: $S \leftarrow V$
6: **while** $S \neq \emptyset$ **do**
7:     **find** $v_m \in S$ with minimum d ▷ find node v_m with minimum distance
8:     **for all** $\{v_m, u\}$ **do**   ▷ update each neighbor distance of v_m
9:         **if** $d_u > d_{v_m} + length(\{u, v_m\})$ **then**
10:            $d_u \leftarrow d_{v_m} + length(\{u, v_m\})$
11:        **end if**
12:    **end for**
13:    $T \leftarrow T \cup \{v_m\}$ ▷ include new node in the shortest path
14:    $S \leftarrow S \setminus \{v_m\}$ ▷ remove new node from searched
15: **end while**

---

are described. The first algorithm due to Dijkstra is inherently sequential; however, the second algorithm by Bellman and Ford is suitable for distributed processing.

7.2.1 Dijkstra's Algorithm

The SSSP algorithm proposed by Dijkstra [3] (*Dijkstra_SSSP*) computes all shortest paths from a single node. It can be applied by each node of the network graph to find All-Pairs Shortest-Paths (APSP) of the network. The idea of this algorithm is to start from a source node s and include iteratively in the route the nodes with the lowest costs from s. As shown in Algorithm 7.1, S is the set of nodes for which shortest paths have not been found, and d_u for node u is the shortest known distance from the source node s to node u. The algorithm starts by setting $S = V$ and $d_u = \infty$ for each node u except the source node s, which has $d_s = 0$. At each iteration, the vertex v that has the minimum distance value to the source is deleted from S, and each neighbor u of v is investigated to find if a path through v provides a shorter path to s than the current distance d_u.

An example execution of *Dijkstra_SSSP* is shown in a directed graph G in Fig. 7.1, where node a is the source node from which the shortest paths are to be computed. The node with the lowest distance is node a itself as all others have infinite distances initially. Nodes b and e that are neighbors of a are marked with distances 8 and 2, respectively, and a is added to T, removed from S in the first iteration of the loop. In the second iteration of the *for* loop, node b has the lowest distance in S, its neighbors e and c are marked with distances 3 and 8, and the previous distance 8 of node e is changed to 3 in this iteration. The algorithm proceeds similarly, adding a vertex to the already decided list of vertices T in each iteration, and finally, $T = \{a, b, e, d, c\}$ in sequence is obtained.

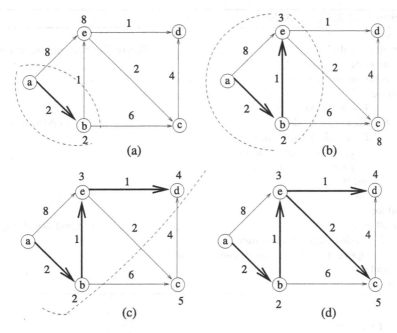

Fig. 7.1 *Dijkstra_SSSP* execution example. (**a**) The graph G. (**b**) First iteration. (**c**) Second iteration. (**c**) Final iteration

The time complexity of *Dijkstra_SSSP* is $O(n^2)$, considering the two nested loops. Using suitable data structures, the time can be reduced to $O(n \log n)$ steps, and for the APSP problem, $O(n^3)$ (or $O(n^2 \log n)$) operations are needed as the algorithm will be executed for each node. It should be noted that to calculate the APSP routes in a distributed system, all the nodes should have the network connectivity information such as the adjacency matrix of the graph. For this reason, *Dijkstra_SSSP* is not convenient for distributed processing.

7.2.2 Bellman–Ford Algorithm

Dijkstra_SSSP provides wrong answers for graphs with negative weight cycles. The Bellman–Ford SSSP algorithm (*BellFord_SSSP*) [1] solves this problem using dynamic programming and works for negative weight edges and cycles by iteratively updating distances from the source node using the results obtained from the previous iteration. Given a weighted graph $G(V, E, w)$ with source s, *BellFord_SSSP* outputs the shortest paths from s and their weights if there is no negative weight cycle, and it produces no answer if there is a negative weight cycle as shown in Algorithm 7.2. *BellFord_SSSP* works with directed and undirected graphs.

Algorithm 7.2 *BellFord_SSSP*

1: $d_s \leftarrow 0$
2: **for all** $i \neq s$ **do**   ▷ initialize distances and predecessors
3:     $d_i \leftarrow \infty$
4:     $predecessor(u) \leftarrow \varnothing$
5: **end for**
6: **for** $k = 1$ **to** $n - 1$ **do**
7:     **for all** $\{u, v\} \in E$ **do**   ▷ update distances
8:         **if** $d_u > d_v + length(u, v)$ **then**
9:             $d_u \leftarrow d_v + length(u, v)$
10:             $predecessor(u) \leftarrow v$
11:         **end if**
12:     **end for**
13: **end for**
14: **for all** $(u, v) \in E$ **do**   ▷ report negative cycle
15:     **if** $d_u + length(u, v) > d_v$ **then**
16:         **report** "*Graph contains a negative cycle*"
17:     **end if**
18: **end for**

7.2.2.1 Example

An example execution of *BelFord_SSSP* is depicted in the same directed graph G in Fig. 7.2, where node a is the source node from which the shortest paths are to be computed. Each edge is checked in each iteration, and if the new distance calculated with the relaxation is smaller, the node is labeled with the new distance, and its predecessor is marked. In the first iteration in (b), nodes b and e are marked with 2 and 8 distances, respectively, and all other nodes remain with infinite distances to the source node a. In the second iteration, one-hop distances from a are propagated to two-hop neighbors of a, and if these new paths yield lower cost paths, routes are modified. The final routes shown in (d) are the same as found by *Dijkstra_SSSP*.

As there will be $|E|$ edges checking at each iteration and to consider the longest path of $n - 1$ hops, the outer *for* loop needs to be executed $n - 1$ time, the total time complexity of *BelFord_SSSP* is $|E|(n - 1)$, that is, $O(nm)$. Since each edge checking can be performed independently, *BelFord_SSSP* is more suitable for distributed processing as we will see in the following sections.

7.2.3 All-Pairs Shortest-Paths Routing Algorithm

The Floyd–Warshall Algorithm (*FW_APSP*) finds APSP routes in a graph G and works with positive or negative weights on edges. It uses dynamic programming by comparing all possible paths between each pair of nodes in G by incrementally improving the shortest path between them until the estimate is optimal. It starts with

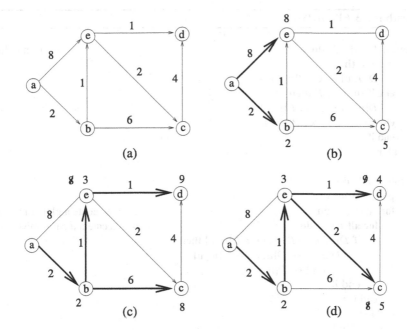

Fig. 7.2 *BellFord_SSSP* execution example

a path that consists of a single edge and iteratively computes paths by increasing the set of intermediate nodes by adding a pivot node to the intermediate nodes. It assumes that all cycles in the graph have positive total weights. For $S \subseteq V$, a path $\langle u_0, \ldots, u_k \rangle$ is an S-path if all internal nodes belong to S, that is, $u_i \in S$ for all $0 < i < k$. For $\{u, v\} \in V$, the S-distance $d^s(u, v)$ is the lowest weight of an S-path [6]. If $S' = S \cup \{w\}$, then a simple path from u to v is either a shortest S-path from u to v or a shortest path from u to w added to the shortest path from w to u, whichever is shorter. Therefore,

$$d^{S'}(u, v) = min(d^S(u, v), d^S(u, w) + d^S(w, v)). \tag{7.1}$$

The algorithm maintains a matrix $D[n, n]$ that shows the current distance between two nodes u and v. There is also the matrix $P[n, n]$ that contains the first node on the current shortest path from u to v. For every $w \in V$, a check is done whether for every node u and $v \in V$, the distance through the intermediate node w will yield a shorter distance than the current one. If this is valid, a new distance is calculated through w and inserted in D. The path through w is also shown by assigning w in the matrix P as shown in Algorithm 7.3. As the inner loop is executed n^2 times and the outer loop n times, the time complexity of *FW_APSP* is $O(n^3)$. Initialization also takes n^2 steps.

Algorithm 7.3 *FW_APSP*

```
 1:  S ← ∅
 2:  for all {u, v} ∈ V do                                          ▷ initialize
 3:      if u = v then
 4:          D[u, v] ← ∅, P[u, v] ←⊥
 5:      else if {u, v} ∈ E then
 6:          D[u, v] ← w_uv, P[u, v] ← v
 7:      else D[u, v] ← ∞, P[u, v] ←⊥
 8:      end if
 9:  end for
10:
11:  while S ≠ V do
12:      pick w from V \ S
13:      for all u ∈ V do                                  ▷ Execute a global w-pivot
14:          for all v ∈ V do                              ▷ Execute a local w-pivot at u
15:              if D[u, w] + D[w, v] < D[u, v] then
16:                  D[u, v] ← D[u, w] + D[w, v]
17:                  P[u, v] ← P[u, w]
18:              end if
19:          end for
20:      end for
21:      S ← S ∪ {w}
22:  end while
```

7.2.3.1 Example

Figure 7.3 shows the iterative operation of *FW_APSP* on the sample network where after five iterations with pivots in sequence as a, b, c, d, and e, all the shortest distances between every pair of nodes are determined, and the changed distances are shown in bold. The P matrix shown in the final form in (d) displays the first node on the shortest path from node u to node v.

7.3 The Distributed Floyd–Warshall Algorithm

DistFW_APSP chooses a pivot node w as an intermediate node to calculate distances, and the rest of the computation is local. In order to provide a distributed algorithm based on *FW_APSP*, the choice of w should be done globally. In our first and incomplete attempt to design such an algorithm, each node holds a local vector $D_u[n]$, where $D_u[v]$ is the current shortest distance from node u to node v. We also have a local vector $P_u[n]$, where $P_u[v]$ shows the first node along the shortest path from u to v. Operation of this algorithm called *DistFW_APSP* is similar to *FW_APSP* with the difference of broadcasting of the vector D_w by a single node to all nodes. All nodes should decide on the same node w, and this can be done by providing unique labels to all nodes of the graph and proceeding in lexicographical

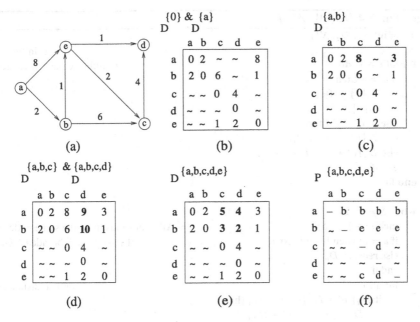

Fig. 7.3 *FW_APSP* execution example

order. Any node that receives the broadcast D_w updates its D_u as in *FW_APSP*. It should be noted that all node identifiers should be known to all nodes of the graph initially and the graph should not contain any negative cycles as shown in Algorithm 7.4, similar to [6].

The algorithm terminates after n iterations at each node. We will defer computing the message complexity of this algorithm as we have not decided on how to broadcast D_w to all nodes.

7.4 Toueg's Algorithm

Toueg provided a distributed algorithm based on *FW_APSP* to find shortest paths in a network without broadcasting the vector D_w to all nodes [7]. Let $T_w = (V_w, E_w)$ be a rooted tree toward w stored in Nb_u at the beginning of iteration that adds pivot w. Toueg observed that node u which has $D_u[w] = \infty$ at the start of the w-pivot round does not change its vectors in this round. Therefore, the only nodes that may receive D_w to update their tables are the members of the current T_w, which means that we only need to broadcast D_w over the edges of T_w. All the nodes in T_w should know their children in T_w to provide the broadcast; however, only the children know their parents by P_w vector from the previous pivot operation. In *Toueg_APSP* algorithm, each node u sends a *child* message to its parent neighbor and *nonchild* to all other neighbors, so that parents are notified about their children.

Algorithm 7.4 *DistFW_APSP*

1: **set of int** S_u, D_u, Nb_u
2: $S_u \leftarrow \varnothing$ ▷ initialize
3: **for all** $v \in V$ **do**   ▷ initialize data structures
4:     **if** $u = v$ **then**
5:         $D_u[v] \leftarrow 0$
6:         $P_u[v] \leftarrow \perp$
7:     **else if** $uv \in E$ **then**
8:         $D_u[v] \leftarrow w_{uv}, P_u[v] \leftarrow v$
9:     **else** $D_u[v] \leftarrow \infty, P_u[v] \leftarrow \perp$
10:     **end if**
11: **end for**
12:
13: **while** $S \neq V$ **do**
14:     **pick** $w \in \{V \setminus S\}$ ▷ All nodes must pick the same node w
15:     **if** $u = w$ **then broadcast** D_w ▷ if i am picked, broadcast D_w
16:     **else receive** D_w
17:     **end if**
18:     **for all** $v \in V$ **do**   ▷ update distances
19:         **if** $D_u[w] + D_w[v] < D_u[v]$ **then**
20:             $D_u[v] \leftarrow D_u[w] + D_w[v]$
21:             $P_u[v] \leftarrow P_u[w]$
22:         **end if**
23:     **end for**
24:     $S \leftarrow S \cup \{w\}$
25: **end while**

The node u then waits for its own *child* and *nonchild* messages to complete the building of T_w tree. Once this is accomplished, the next phase of the algorithm involves nodes that have $D_w \neq 0$, as these may change their D_w values similar to *FW_APSP* as shown in Algorithm 7.5.

7.4.1 Analysis

Theorem 7.1 *The time complexity of Toueg_APSP is $O(n^2)$, and its message complexity is $O(nm)$. Its bit complexity is $O(n^3 \log n)$, and it requires $O(n \log n)$ bits of storage per node.*

Proof The main loop is executed n times, and it contains a loop with n iterations resulting in $O(n^2)$ time. Each edge is traversed three times by *child*, *nonchild*, and at most once by D_w messages in the w-pivot round for a total of at most $3n$ messages per edge. The total number of messages traversed is therefore $O(nm)$. Each D_w message has $O(n \log n)$ bits, and each *child* and *nonchild* message has $O(\log n)$ bits. As there will be $O(n^2)$ D_w messages and $2mn$ *child* and *notchild* messages, the

Algorithm 7.5 *Toueg_APSP*

1: **set of int** S_u, D_u, Nb_u
2: $S_u \leftarrow \varnothing$ ▷ Initialize
3: **for all** $v \in V$ **do**
4:     **if** $u = v$ **then**
5:         $D_u[v] \leftarrow 0, Nb_u[v] \leftarrow \perp$
6:     **else if** $\{u, v\} \in E$ **then**
7:         $D_u[v] \leftarrow w_{uv}, Nb_u[v] \leftarrow v$ ▷ initialize neighbor costs
8:     **else** $D_u[v] \leftarrow \infty, Nb_u[v] \leftarrow \perp$ ▷ all other nodes are undefined
9:     **end if**
10: **end for**
11: **while** $S_u \neq V$ **do** ▷ Build T_w
12:     **pick** w from $V \setminus S_u$ ▷ choose a global pivot
13:     **for all** $x \in \Gamma(u)$ **do**
14:         **if** $Nb_u[w] = x$ **then send** *child*(w) to x ▷ send *child* to parent
15:         **else send** *nonchild*(w) to x ▷ send *nonchild* to other neighbors
16:         **end if**
17:     **end for**
18:     $n\_recvd \leftarrow 0$
19:     **while** $n\_recvd < |\Gamma(u)|$ **do** ▷ receive status messages from all neighbors
20:         **receive** a *child*(w) or *nonchild*(w)
21:         $n\_recvd \leftarrow n\_recvd + 1$
22:     **end while**
23:     **if** $D_u[w] < \infty$ **then** ▷ Only nodes on T_w execute this part
24:         **if** $u \neq w$ **then**
25:             **receive** D_w from $Nb_u[w]$ ▷ receive distance values from pivot w
26:         **end if**
27:         **for all** $x \in \Gamma(u)$ that sent *child*(w) **do send** D_w to x ▷ send pivot distances to children
28:         **end for**
29:         **for all** $v \in V$ **do** ▷ update distance values
30:             **if** $D_u[w] + D[v] < D_u[v]$ **then**
31:                 $D_u[v] \leftarrow D_u[w] + D[v]$
32:                 $Nb_u[v] \leftarrow Nb_u[w]$
33:             **end if**
34:         **end for**
35:     **end if**
36:     $S_u \leftarrow S_u \cup \{w\}$
37: **end while**

---

total number of bits communicated will be $O(n^2 n \log n + 2n \log n) = O(n^3 \log n)$. The D_u and NB_u tables at each node require $O(n \log n)$ bits. $\qquad \square$

A problem with *Toueg_APSP* is that it requires the global knowledge about node identifiers to be able to decide on the same pivot w at each round. This would require a prior execution of another algorithm to broadcast identifiers.

Algorithm 7.6 *DBF_APSP*

 1: **message types** *round*, *update*
 2: **int** $i, j, my\_dist, dist$
 3: **set of int** *received* $\leftarrow \varnothing$
 4: **boolean** *round_over* $\leftarrow$ *false*, *round_recvd* $\leftarrow$ *false*
 5: **while** $\neg round\_over$ **do**   ▷ A single round executed by each node except the source
 6:     **receive** $msg(j)$
 7:     **case** $msg(j).type$ **of**
 8:         $\underline{round(k)}$:   **send** $update(k, my\_dist)$ to Γ_i
 9:                                   $round\_recvd \leftarrow true$
10:         $\underline{update(k, dist)}$: $received \leftarrow received \cup \{j\}$
11:                                   **if** $received = \Gamma_i \wedge round\_recvd$ **then**
12:                                       **for all** $j \in \Gamma(i)$ **do**
13:   **if** $my\_dist > (dist + w_{ij})$
14: $my\_dist \leftarrow dist + w_{ij}$
15: $parent \leftarrow j$
16:                                   $round\_over \leftarrow true$
17: **end while**

7.5 Synchronous Distributed Bellman–Ford Algorithm

The synchronous distributed routing algorithm based on *BelFord_SSSP*, which we will call *DBF_APSP*, is initiated by a single node and works in rounds. In each round, each node sends its current distance from the source to all of its neighbors. It then collects all the distance values from all its neighbors, decides on the shortest distance via its neighbor to the source, and marks this neighbor as its parent, shown as a single round in Algorithm 7.6. The root node initiates each round, and there should be $n - 1$ rounds for the distance values to reach the longest path in the graph. For this reason, the root node should be aware of the number of nodes in the network to decide on the number of rounds to be executed. The synchronization messages are not shown, and it may be assumed that a spanning tree has already been formed to allow sending synchronization messages *round* and *upcast*. The boolean *round_recvd* variable ensures that *round* message is received before updating distances as it is possible to receive the *update* messages from all neighbors before receiving the *round* message in a round. The operation of this algorithm on the sample network would provide the same tree as the *BellFord_SSSP* example.

7.5.1 Analysis

Theorem 7.2 *The time complexity of DBF_APSP is $O(n)$, and its message complexity is $O(nm)$.*

Proof As the root executes a total of $n - 1$ rounds, which is the longest path in the graph, Time($BelFord\_Synch$) $= O(n)$. Each edge will be traversed exactly twice,

---

Algorithm 7.7 *CM_APSP*

1: **int** *parent* $\leftarrow \perp$, *my_dist* $\leftarrow \infty$, *n_acks* $\leftarrow 0$, *n_weighting* $\leftarrow 0$
2: **boolean** *finished* $\leftarrow$ false
3: **message types** *update*, *ack*
4: **if** $i = root$ **then**   ▷ root sends its distances to all its neighbors
5:     **send** $dist(w_{i,j})$ to $\Gamma(i)$
6:     $n\_weighting \leftarrow |\Gamma(i)|$
7: **end if**
8: **while** ¬*finished* **do**
9:     **receive** $msg(j)$
10:     **case** $msg(j).type$ **of**
11:          $update(dist)$:   **if** $my\_dist > dist$ **then**   ▷ update distances
12:                         $my\_dist \leftarrow dist$
13:                         **if** $parent \neq j$ **then** $parent \leftarrow j$
14:                         **send** $update(my\_dist + w_{i,j})$ to $\Gamma(i) \setminus \{parent\}$
15:                         $n\_weighting \leftarrow n\_weighting + |\Gamma(i)| - 1$
16:         ack: $n\_acks \leftarrow n\_acks + 1$
17:                         **if** $n\_acks = n\_weighting$ **then**   ▷ all *acks* received
18:                         **send** *ack* to j
19:                         *finished* $\leftarrow$ *true*
20: **end while**

---

once in each direction by the *update* messages in each round for a total of $2m$ messages per round, resulting in $\text{Msg}(BelFord\_Synch) = O(nm)$. If synchronization is performed by a protocol that uses the synchronization messages *round* and *upcast* over a spanning tree T, there will be additional $O(2(n-1))$ synchronization messages per round, resulting in a total of $O(mn + n^2)$ messages.   □

7.6 Chandy–Misra Algorithm

The Chandy–Misra algorithm (*CM_APSP*) [2] is based on the asynchronous execution of *BelFord_SSSP* with augmented termination. The working principle is the same relaxation method we have seen before. However, there is no central synchronization as in the synchronous version of the algorithm. The essential component of the algorithm is that when the distance of a node to the source is changed, it sends the *update* message to its neighbors and adds the count of neighbors to the neighbors it was already waiting as shown in lines 14–15 of Algorithm 7.7, deferring sending of the *ack* message to the parent until it has received acknowledgements from all of the nodes that it is waiting.

As a node will receive its optimum distance at most at $O(n)$ hops away from the root, the time complexity of *CM_APSP* is $O(n)$. However, the number of messages is exponentially bounded. If all link costs are assumed to be equal, the shortest path from a single source can be calculated using $O(n^2)$ messages per edge and $O(mn^2)$ messages in total each with $O(\log n)$ bits [6].

7.7　Routing Protocols

In this section, we will review two important routing protocols that have been used in the Internet, the *Link State Protocol* and the *Distance Vector Protocol*.

7.7.1　Link State Protocol

Link State Protocol relies on flooding the local connectivity of each node to the network so that every node has a complete view of the network after which it can run *Dijkstra_SSSP* algorithm to compute the shortest paths [5]. Periodically, each node sends its Link State Packet (LSP), which includes its identifier, list of neighbors and costs (delays to each neighbor), sequence number, and a special field called *Time to Live* (TTL) to all of its neighbors. Upon receiving an LSP, it checks whether this is the most recent one it has seen, and if so, sends it to all of its neighbors. This operation can be specified as follows:

1. Periodically send LSP to all neighbors.
2. When an LSP is received that is recent:
   a. Use *Dijkstra_SSSP* to compute the distances to all other nodes
   b. Store ⟨destination, next hop⟩ pair in the forwarding table.
3. If LSP is not recent, discard.

The link state protocol is used in OSPF protocol of Internet Protocol. Its main drawback is the size of the local storage required at each routing node as the whole network table is to be stored.

7.7.2　Distance Vector Protocol

The Distance Vector Routing protocol (DVR) is based on *BelFord_SSSP* algorithm and was used in the ARPANET up to 1980 [4]. The following are the main differences between the *BelFord_SSSP* and the DVR protocol:

- DVR protocol is run in a continuous loop to compute new routes for dynamic networks.
- Each node i holds an array $length[1..n]$, where entry $length[j]$ shows the distance from node j to i.
- *length* is included in *update* messages and shows the distances of the sender.
- Node i that receives an *update* message from j compares each *length* entry in the message by its own *length* values and updates its array to give the shortest paths.
- Each node i also has an array $parent[1..n]$ as the routing table, where $parent[j]$ has an entry for the next node to route the packet destined to j.

Table 7.1 Comparison of distributed routing algorithms

Algorithm	Description	Time Comp.	Msg Comp.
Toueg_APSP	Broadcasts table over tree	$O(n^2)$	$O(nm)$
DBF_APSP	Needs synchronization	$O(n)$	$O(nm)$
CM_APSP	Asynchronous with termination detection	$O(n^2)$	$O(n^2m)$

Fig. 7.4 Example graph for Exercise 1

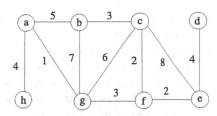

The processes exchange their distance vectors periodically, and if a message does not arrive within the period, a default value is assumed, and the next round is started. A known problem with the DVR protocol is the *count-to-infinity* problem, which is encountered when a link to an isolated node breaks and the nodes start to increase their distances to the failed node.

7.8 Chapter Notes

We have analyzed few distributed routing algorithms and two protocols based on fundamental sequential algorithms of Dijkstra and Bellman–Ford. Table 7.1 summarizes the distributed algorithms. *DBF_APSP* algorithm has low time complexity, however, requires synchronization at middleware level by a synchronizer or by a network wide protocol, both of which result in more message overhead.

Chandy–Misra Algorithm, although simple, has exponential message complexity for a general weighted graph. Toueg's Algorithm requires identifiers to be available globally prior to execution, which necessitates running of another algorithm to provide these identifiers. Link State Protocol does not have the count-to-infinity problem that the Distance Vector Protocol has, but it has significant memory requirements.

Routing remains a well studied but still a fundamental problem in computer networks. We will see routing in ad hoc wireless networks in Chap. 8, where there will be the mobility and energy levels of the nodes to be considered additionally.

7.8.1 Exercises

1. Show the execution of *Dijkstra_SSSP* and *BelFord_SSSP* algorithms in the sample graph of Fig. 7.4 to find shortest routes to vertex a.

Fig. 7.5 Example graph for
Exercise 2

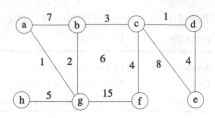

2. Show a possible execution of *DBF_APSP* in the sample graph of Fig. 7.5 to find routes to vertex *a*.

3. Modify *CM_APSP* algorithm so that storage at a node of the identifiers of nodes in its subtree is provided. Write the pseudocode of a distributed algorithm where a search by the message *query(node_id)* is initiated by any node and a node in the subtree returns its path to the searching node. Work out the time and message complexities of this algorithm.

References

1. Bellman R (1958) On a routing problem. Q Appl Math 16:87–90
2. Chandy KM, Misra J (1982) Distributed computation on graphs: shortest path algorithms. Commun ACM 11:833
3. Dijkstra EW (1959) A note on two problems in connexion with graphs. Numer Math 1:269–271
4. Hedrick C (1988) RFC1058—routing information protocol. Internet Engineering Task Force
5. McQuillan JM, Richer I, Rosen EC (1978) ARPANet routing algorithm improvements. BBN report no. 3803, Cambridge
6. Tel G (2000) Introduction to distributed algorithms, 2nd edn. Cambridge University Press, Cambridge
7. Toueg S (1980) An all-pairs shortest-path distributed algorithm. Technical report RC 8327, IBM TJ Watson Research Center, Yorktown Heights, NY 10598

Chapter 8
Self-Stabilization

Abstract A distributed algorithm is *self-stabilizing* if starting from any state, it eventually reaches an allowed (legal) state. A self-stabilizing system running self-stabilizing algorithms recovers from faults and, once recovered, stays recovered. Self-stabilizing algorithms typically run in background and never stop. In this chapter, we review basic self-stabilization concepts and analyze BFS and DFS self-stabilizing algorithms.

8.1 Introduction

As proposed by Dijkstra, a system is self-stabilizing when, regardless of its initial state, it is guaranteed to arrive at a legal state in a finite number of steps [7]. When in the legal state, it only moves to another legal state in the absence of faults. A legal state is defined as the state in which the system has the desired property required by the application and is *error-free*. A distributed system running a self-stabilizing algorithm is called a *self-stabilizing system*. Self-stabilizing systems aim to provide fault tolerance by recovering from faults in a bounded time without any external intervention.

Every computing element of a self-stabilizing system works as a state machine and has a local state determined by the values of its local variables. Each node reads its input variables and based on its current state and its inputs, it may perform a state transition and produce some output. The global state of the system is the union of all local states as in a general distributed system. The global state can be either *legal* (legitimate) or *illegal* (illegitimate). Every node of the system executes repeatedly rules as

$$(label)[guard] : \langle action \rangle$$

A *guard* is a boolean expression of the variables of a node and its neighbors, and if the guard of a *rule* is evaluated as true, that rule is *enabled*, and a node with at least one rule enabled becomes *privileged* so that the related *action* may be executed.

A *move* of a node is nondeterministic execution of an enabled rule in that node. A privileged node may be chosen to execute as decided by a scheduler. An enabled rule as a result of the move of a node computes new values for the local variables. A new move may not be started until the previous move is completed, and hence

K. Erciyes, *Distributed Graph Algorithms for Computer Networks*,
Computer Communications and Networks, DOI 10.1007/978-1-4471-5173-9_8,
© Springer-Verlag London 2013

the moves are *atomic*. In other words, a move consisting of reading the state of neighbors and modifying the local state cannot be interrupted.

Two important requirements from a self-stabilizing system are *closure* and *convergence* properties, and the general requirement to prove the correctness of a self-stabilizing algorithm is to show that it has these properties. The closure property means that the system only makes legal moves between two states. The convergence property ensures that starting from any state, the system reaches a legal state after a finite number of state transitions. This property eliminates the need for termination detection in a self-stabilizing system.

8.2 Models

Dijkstra [7] proposed the concept of a *central daemon* in which a special process selects one of the privileged nodes to make the next move, which is called the *serial execution*. Proving the correctness of serial execution is simpler; however, this model is not convenient for distributed processing as it does not allow concurrent processing. In the *distributed daemon* model, each node decides independently for its next move.

In the *restricted parallelism model*, there may be specific restrictions on the set of nodes that may execute at each step, whereas all the enabled processes may execute in the *maximum parallelism* model.

An iterative method to prove self-stabilizing algorithms, which is difficult in many cases, involves assuming a central daemon first. When the correctness is verified in this case, the assumption is removed, and it is checked whether the system still works correctly. If not, the algorithm may need to be extended. For example, it was shown that self-stabilizing algorithms due to Dijkstra, which work with a central daemon, can work correctly with a distributed daemon [5].

8.2.1 Anonymous or Identifier-Based Networks

Nodes of a self-stabilizing system can have identifiers that are unique, in which case the network is called *id-based*, or no identifiers, where it is called *anonymous*. In general, designing algorithms for *id*-based networks is much simpler than anonymous networks as symmetry breaking can be accomplished by the use of identifiers. In some cases, there are no deterministic algorithms for anonymous networks. Real networks are *id*-based as each node has an IP number and an Ethernet address.

8.2.2 Deterministic, Randomized, or Probabilistic Algorithms

A *randomized algorithm* employs some randomness during its execution and generally uses random inputs to achieve the required result. The performance of the algorithm depends on the value of the random input so that the output and/or the expected running time are random variables. It is generally required that a randomized algorithm should have favorable performance in the average case of the randomly selected inputs. In general, these algorithms are used to break symmetries in anonymous self-stabilizing networks. A *probabilistic algorithm* also depends on a random input, but it is not guaranteed to provide a correct result by these algorithms. However, probabilistic algorithms may be the only choice in some situations. A *randomized self-stabilizing system* reaches a stable state in a bounded, expected number of moves. A self-stabilizing algorithm that is neither randomized nor probabilistic is called *deterministic*.

8.3 Dijkstra's Self-Stabilizing Mutual Exclusion Algorithm

Dijkstra [7] proposed three self-stabilizing mutual exclusion algorithms for a token ring. The aim of these algorithms is to reach a stable configuration of the system so that there is a single process that is privileged executing a critical section at any time. In the first algorithm, there are N processes $p_0, \ldots, p_{N-1}$ that form a unidirectional ring. Each process has a variable $x_i \in \{0, \ldots, K - 1\}$, where $K \geq N$. Any process other than the root p_0 becomes privileged if $x_i \neq x_{i-1}$, and p_0 becomes privileged if $x_0 = x_{N-1}$. The algorithm applies two rules:

- **if** $x_i \neq x_{i-1}$ **then** $x_i = x_{i-1}$ for $0 < i < N$ {any process other than the root}
- **if** $x_0 = x_{i-1}$ **then** $x_0 = (x_0 + 1) \bmod K$ {the root process}

When the algorithm stabilizes, there will be only one node changing state; therefore, mutual exclusion principle will be provided. Figure 8.1 displays the execution of this algorithm for five processes $p_0, \ldots, p_4$, where p_0 is the root. The privileged nodes at each iteration are shown by double circles, and the enabled node is shown by a triple circle. As can be seen, stability is reached in (f), where there is only one node enabled after which there will only be one privileged node in sequence starting by the node p_1 and the mutual exclusion is provided.

8.4 BFS Tree Construction

In this section, we describe two self-stabilizing algorithms that construct BFS trees of the network. The first algorithm uses a central scheduler, and the second one works with a distributed scheduler.

8.4.1 Dolev, Israeli, and Moran Algorithm

Dolev, Israeli, and Moran [8] proposed a self-stabilizing BFS spanning-tree construction algorithm (*DIM_BFS*) for semi-uniform systems (systems with a distinguished node) with a central daemon under read/write atomicity. Each node has a pointer to one if its incoming edges and an integer showing the distance in hops to

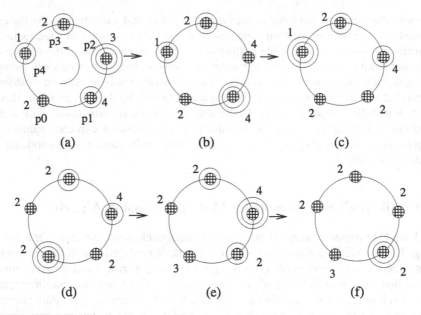

Fig. 8.1 Dijkstra's mutual exclusion algorithm example

Algorithm 8.1 *DIM_BFS*

1: **int** *dist*, *parent*
2: **if** $i = root$ **then**
3:       **while** *true* **do**
4:             $dist \leftarrow 0, parent \leftarrow -1$
5:       **end while**
6: **else**
7:       **while** *true* **do**
8:             $\forall u \in \Gamma(i)$ **receive** $dist(u)$
9:             $j \leftarrow min\{\bigcup_i dist(i)\}$ ▷ find node j that is minimum distance to the root
10:           $parent \leftarrow j$ ▷ mark j as the parent
11:           $dist \leftarrow dist_j + 1$ ▷ update *dist*
12:       **end while**
13: **end if**

---

the root of the tree, which is a distinct node that always sends a value of 0 periodically. The nodes periodically exchange their distance values (*dist*) with each other, and after a node reads the distance values of all neighbors, it decides the neighbor with minimum distance as its new parent. It then writes its own distance into its output registers as $dist + 1$ as shown in Algorithm 8.1.

The algorithm is started by the root that writes 0 to all its output registers. Immediate neighbors of the root will select it as their parent consequently, and this will not change. After $O(d)$ moves, where d is the diameter of the network, the BFS tree will be formed and stabilized.

8.4.2 Afek, Kutten, and Yung Algorithm

Afek, Kutten, and Yung algorithm [2] (*AKY_BFS*) builds a BFS tree in the read/write atomicity model in a uniform system without a distinguished root process. This algorithm assumes that all nodes have globally unique identifiers that can be totally ordered, and the node with the largest identifier eventually becomes the root of the tree.

Each node has a pointer showing its current parent, a distance variable, and the identifier of the root of the tree. Initially, each node tries to construct a spanning tree rooted at itself. During periodical exchange of information with the neighbors, if a node finds that it has the largest identifier, it becomes the root of its local tree. If a node detects that there is a tree that a neighbor belongs with a root that has a larger identifier, it joins the neighbor tree with a *join* message and by receiving a *grant* message in acknowledgement. When the BFS tree is finally constructed, there is one root that has the largest identifier; each node points to this root by its parent and the *dist* value at the node is 1 more than the distance of its parent to the root.

The local variables at node i are its neighborhood (N_i), its parent node (P_i), its root node (R_i), and its distance to the root node (D_i). The algorithm checks two conditions to determine whether a node is in a legal state or not:

1. A : $[(R_i = i) \wedge (P_i = i)] \vee [(R_i > i) \wedge (P_i \in N_i) \wedge (R_i = R_{\text{parent}}) \wedge (Distance = D_{\text{parent}} + 1 > 0)]$
2. B : $[A \wedge (R_i \geq max(R_N(i)))]$

Informally, if i is the root node, Condition A is satisfied. Otherwise, if its root R_i is larger than its identity and the same as its parent root and its parent is one of its neighbors and its distance to the root is 1 bigger than its parent's distance, Condition A is again satisfied. For Condition B to be true, Condition A should be true, and the root of node i should be greatest among all its neighbors. When both conditions are true, node i is in legal state. The following shows the actions as related to these guards:

- $A \wedge B$: Node i is in legal state.
- $A \wedge B'$: Node i decides to join the tree with largest root identity and sends a request message to the related neighbor.
- $A' \wedge B'$: Node i is in illegal state.

The algorithm also provides other actions for fault tolerance. *AKY_BFS* requires $O(n^2)$ rounds to stabilize [9].

8.5   Self-Stabilizing DFS

Collin and Dolev [6] proposed a semi-uniform spanning-tree algorithm (*CD_DFS*) under a central daemon and read/write atomicity to construct a DFS tree. Key to the operation of the algorithm is the ordering of the outgoing links of each node. The

---

Algorithm 8.2 *CD_DFS*

```
 1: if i = root then
 2:     while true do
 3:         path_i ← ⊥
 4:     end while
 5: else
 6:     while true do
 7:         for j = 1 to |Γ(i)| do
 8:             read_path_j ← read(path_j)
 9:         end for
10:         write path_i ← min{|read_path_j ∘ α_j(i)|_N, such that 1 ≤ j ≤ |Γ(i)|}
11:     end while
12: end if
```

DFS tree is formed by traversing the network graph by continuously selecting the smallest outgoing edge. The root repeatedly writes the *empty path* ($\perp$) to its output registers. All other nodes read the paths of their neighbors to the root and choose the lexicographically minimal of these paths. The choice of smallest link identifiers eventually results in a DFS tree, and each node has a path directed toward the root.

A node of the network is denoted by $P(i)$, and each node $P(i)$ orders its edges by some arbitrary ordering $\alpha(i)$. For any edge $e = (P_i, P_j)$, $\alpha_i(j)$ is defined as the *edge index* of e according to $\alpha(i)$ and $\alpha_j(i)$ similarly according to $\alpha(j)$. Each node has a variable called $path_i$ and reads the path $path_j$ from node j and concatenates $path_j$ with $\alpha_i(j)$. Having formed the concatenated paths from all neighbors, the node $P(i)$ chooses the minimal path as the new value for $path_i$ and writes this path to its output registers as shown in Algorithm 8.2, where N is an upper bound of the number of nodes in the graph.

The memory requirements for the DFS algorithm is $O(n \log K)$ bits, where K is an upper bound of the maximum degree of a node. The time complexity is $O(ndK)$ rounds, where d is the diameter of the network graph [6].

8.6 Chapter Notes

We have reviewed briefly the basic concepts of self-stabilizing systems and showed some fundamental algorithms for BFS and DFS tree construction. These algorithms have many application areas such as mutual exclusion and network protocol design. Arora and Gouda [4] provided a self-stabilizing BFS spanning tree algorithm for the composite atomicity model under a central daemon with unique identifiers and the node with maximum identifier eventually acting as the root of the system; however, their algorithm needs a bound N on the number n of nodes in the network to work correctly. The number of rounds is $O(N^2)$, which can be much larger than $O(n^2)$ [9]. Herman [10] also presented an algorithm that constructs a DFS tree but uses composite atomicity. More recently, Afek and Bremler [1] proposed an al-

Fig. 8.2 Example graph for
Exercise 3

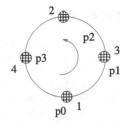

Fig. 8.3 Example graph for
Exercise 4

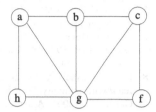

gorithm for systems with unidirectional, bounded capacity message passing links. In
this algorithm, unique identifiers are assumed, and the node with the minimum identifier eventually becomes the root and the number of rounds is $O(n)$. Antonoiu and
Srimani [3] provided a self-stabilizing algorithm to construct an arbitrary spanning
tree of a connected graph.

Self-stabilizing algorithms have significant potential to be used in fault tolerant
distributed systems. However, they have certain limitations. First, many algorithms
using a central daemon are difficult to implement in practice. Second, the system
cannot provide service in an illegitimate state. Some systems may tolerate this in-
terruption, but for some mission critical systems, uninterrupted operation is a necessity. Another difficulty involves proving the correctness and analyzing the complexities of these algorithms, which are usually more complicated than the classical
distributed algorithms.

8.6.1 Exercises

1. Discuss briefly why self-stabilizing algorithms are considered as fault tolerant algorithms. Can they be used in distributed real-time systems where fault-tolerance
   is imperative?
2. Discuss briefly the central and distributed daemon concepts in self-stabilizing
   algorithms.
3. Show a possible execution of Dijkstra's first mutual exclusion algorithm in the
   graph of Fig. 8.2, which shows three processes in a ring with their initial values.
4. Show a possible execution of the Dolev–Israeli–Moran algorithm that builds
   a BFS tree in the sample graph of Fig. 8.3 with the root node a.

References

1. Afek Y, Bremler A (1998) Self-stabilizing unidirectional network algorithms by power supply. Chic J Theor Comput Sci 1998:3
2. Afek Y, Kutten S, Yung M (1991) Memory efficient self-stabilizing protocols for general networks. In: Proc 4th international workshop on distributed algorithms, pp 15–28
3. Antonoiu G, Pradip K, Srimani PK (1995) A self-stabilizing distributed algorithm to construct an arbitrary spanning tree of a connected graph. Comput Math Appl 30:1–7
4. Arora A, Gouda MG (1992) Closure and convergence: a foundation for fault-tolerant computing. In: Proc 22nd international conference on fault-tolerant computing systems
5. Burns JE, Gouda MG, Miller RE (1989) On relaxing interleaving assumptions. In: Proc MCC workshop on self-stabilizing systems
6. Collin Z, Dolev S (1994) Self-stabilizing depth first search. Inf Process Lett 49:297–301
7. Dijkstra EW (1974) Self stabilizing systems in spite of distributed control. Commun ACM 17(11):643–644
8. Dolev S, Israeli A, Moran S (1993) Self-stabilization of dynamic systems assuming only read/write atomicity. Distrib Comput 7:3–16
9. Gartner FC (2003) A survey of self-stabilizing spanning-tree construction algorithms. EPFL technical report
10. Herman T (1991) Adaptivity through distributed convergence. PhD thesis, Department of Computer Science, University of Texas at Austin

Part II
Graph Theoretical Algorithms

Chapter 9
Vertex Coloring

Abstract Vertex coloring is an assignment of colors to the vertices of a graph such that there are no two neighbor nodes having the same color. Vertex coloring has many applications such as task scheduling, register allocation, and channel frequency assignment. In this chapter, we investigate distributed vertex coloring algorithms for arbitrary graphs and trees.

9.1 Introduction

In task scheduling problems, only one task may be assigned to a time slot for execution. The graph consists of vertices representing tasks and edges between vertices for any conflicting tasks that may not execute in the same slot due to sharing of common resources. The minimum number of colors to color this graph is the minimum *makespan*, which is the optimal time for all tasks to finish. Another application is the register allocation where the mostly used data are stored in processor registers during compilation of the program. In this case, the vertices of the graph are the symbolic registers, and there is an edge between two register vertices if they are needed at the same time. Vertex coloring can also be applied to channel frequency assignment in wireless networks such that any adjacent nodes do not use the same frequency to avoid interference. Vertex coloring can be formally defined as follows.

Definition 9.1 (Vertex Coloring) A *vertex coloring* of a graph $G(V, E)$ is the procedure of assigning a color c_v to every vertex $v \in V$ so that c_v is different from any color assigned to the neighbors of v.

In a network with n nodes having unique identifiers, a coloring using n colors is a legal coloring; however, the general requirement for any vertex coloring algorithm is to use as few colors as possible.

Definition 9.2 (Chromatic Number) The chromatic number of a graph G, $\chi(G)$, is the smallest number of colors to color G.

Calculation of $\chi(G)$ is NP-complete [9]. The *clique number* of a graph G, $\omega(G)$, is the number of vertices of the maximum clique that G has. The chromatic number

K. Erciyes, *Distributed Graph Algorithms for Computer Networks*, 107
Computer Communications and Networks, DOI 10.1007/978-1-4471-5173-9_9,
© Springer-Verlag London 2013

of a graph is greater than or equal to its clique number, that is, $\chi(G) \geq \omega(G)$. Brooke' s theorem states that the chromatic number of a graph G is $O(\Delta)$ excluding the complete graph and the odd cycle graph, in which case it is $O(\Delta + 1)$ [3]. A graph G with chromatic number k is called a *k-chromatic graph*, and a graph with $\chi(G) \leq k$ is said to be *k-colorable*. Bipartite graphs are 2-colorable, and the Four-Color theorem states that all planar graphs are 4-colorable [1].

An *edge coloring* of a graph G is a coloring of edges such that there are no adjacent edges of the same color. The *edge chromatic number* χ_e of a graph G is the minimum number of colors that edges of G can be colored and $\chi_e \geq \Delta$. It can be shown that any graph can be edge colored with at most $\Delta + 1$ colors. Therefore, the two classes of graphs are Δ-*edge colored graphs* and $\Delta + 1$ *edge colored graphs*. Determining the edge chromatic number of a graph is NP-complete [16].

A *total coloring* of a graph G is a coloring of G in which all adjacent vertices of G have different colors as in vertex coloring and all incident edges on vertices have different colors as in edge coloring. In this chapter, we first describe two greedy sequential algorithms to color vertices of arbitrary graphs. We then provide algorithms for distributed coloring of arbitrary graphs and algorithms to color trees in the following sections. Finally, two self-stabilizing vertex coloring algorithms are described.

9.2 Sequential Algorithms

Greedy vertex coloring algorithms label the vertices of a graph according to some specific order as $v_1, \ldots, v_n$ and assign the smallest possible color to a vertex v_i that has not been assigned to its neighbors. Such ordering is called a *coloring heuristic*, and one such heuristic is the ordering of the vertices with respect to their degrees. We start by a simple greedy sequential algorithm, called *Seq_Vcol*, to color a graph G by picking uncolored vertices randomly. The *palette* contains a list of colors up to $O(\Delta + 1)$. The *neigh_cols*[n] is an array of set of colors, where *neigh_cols*[u] entry shows the current assigned colors of the neighbors of node u. The algorithm proceeds by assigning the smallest available color to an uncolored vertex u from the *palette* that its neighbors have not been assigned. Once the color c_i is assigned to u, c_i is added to the *neigh_cols*[u] for each neighbor v that is a neighbor of u so that this color is not used in future as shown in Algorithm 9.1.

Figure 9.1(a) displays a graph G with vertex set $V = \{a, b, c, d, e, f\}$ colored by *Seq_Vcol*, where the labels of the vertices denote the colors. The palette consists of colors c_1, c_2, c_3, c_4, c_5 as Δ is 4. The random choice of uncolored vertices is $\{a, c, d, b, e, f\}$ in sequence, and each vertex is colored with the smallest color that its neighbors are not assigned. In (b), the heuristic now is the degrees of vertices in decreasing order, which we call the *highest-degree-first* heuristic, that results in the coloring sequence of $\{b, e, a, d, c, f\}$, where ties are broken by choosing the lexicographically lower vertex. The number of colors used is 4 in (a) and 3 in (b), and color c_5 in *Palette* is not used.

Algorithm 9.1 *Vcol_Seq*

1: **set of int** *Palette* ← {$c_1, c_2, \ldots, c_{\Delta+1}$}; S, *free_cols, neighs_cols*[Δ] ← ∅
2: **int** *colors*[n] ← 0
3: Input $G(V, E)$
4: S ← V
5: **while** S ≠ ∅ **do**
6:     **select** any $u \in S$
7:     *free_cols* ← *Palette* \ *neighs_cols*[u]
8:     **pick** first element $c \in free\_cols$
9:     *colors*[u] ← c
10:    **for all** $v \in \Gamma(u)$ **do**
11:        *neighs_cols*[v] ← *neighs_cols*[v] ∪ {c}
12:    **end for**
13:    S ← S \ {u}
14: **end while**

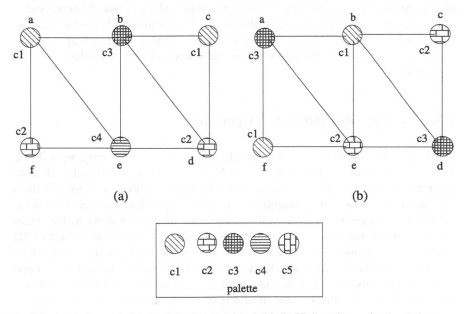

Fig. 9.1 A coloring with *Seq_Vcol*. (**a**) Random heuristic; (**b**) Highest-degree-first heuristic

9.2.1 Analysis

Theorem 9.1 *The time complexity of Seq_Vcol is $\Theta(n)$, and it uses $O(\Delta + 1)$ colors.*

Proof There will be one color assignment per node resulting in the $\Theta(n)$ time complexity. Since each node has a maximum of Δ neighbors, there will always be an available color in the range {$1, 2, \ldots, \Delta+1$}, and hence $O(\Delta+1)$ colors are used. □

9.3 Distributed Coloring Algorithms

In this section, we propose algorithms for arbitrary graphs using the identifiers of nodes, a random algorithm for coloring and two reduction algorithms to reduce the coloring of a colored graph. All of these algorithms work in synchronous rounds. We assume that a spanning tree T is constructed prior to algorithm executions to provide synchronization. Each round is started by a *round* message that arrives at each node at approximately the same time by the broadcasting of the root of T, and any intermediate node that receives this *round* message transfers it to its children. It is possible that a node may receive messages from their neighbors before the *round* message to itself, and control by the boolean variable *round_recvd*, which shows that the *round* message is previously received, provides reception of the messages in any order. The termination of a round is performed by the convergecast *upcast* messages to the root, which can then initiate the next round. The *upcast* messages toward the root and downcast of *round* messages to children are not shown for simplicity. Another issue is that the number of rounds, which is usually expressed in terms of Δ or n, should be known by the root of T beforehand. This restriction may be relaxed if a node that has finished execution upcasts a *finish* message over T, which is convergecast to the root node so that it knows which nodes have finished execution.

9.3.1 The Greedy Distributed Algorithm

In our first attempt to find a distributed algorithm for vertex coloring, we will use a strategy where the highest identifier node in a closed neighborhood decides to color itself with the minimum unused color by their neighbors and informs them by sending a message. Any neighbor node that receives this message deletes the colored higher identifier node from the graph by removing it from the active neighbor list and also removes the color that this neighbor has used from the list of the free colors it may use. Algorithm 9.2 shows the operation of this algorithm, which is called *Rank_Vcol*. Any node that is started by the *round* message in a round checks whether its has the highest identifier among the remaining active neighbors (*curr_neighs*) it has. If this is true, it selects the minimum unused color c from the free colors (*free_cols*) and sends the *color(c)* message to its active neighbors. The active neighbors simply delete the sending neighbor and the color from their lists. For synchronization purposes, the neighbors that do not have the highest identifier among their neighbors simply send the *discard* message so that node u knows that it has received all messages from its neighbors, and the round is over.

In Fig. 9.2, a sample network of eight nodes with identifiers $1, \ldots, 8$ is colored with the *Rank_Vcol* algorithm. In the first round shown in (a), nodes 8 and 7 are the highest identifier nodes in their neighborhoods as shown by double circles and select the minimum unused color in their free colors, which is c_1, and send the $color(c_1)$ message to their lower identifier neighbors, which exclude c_1 from their free colors list and delete these nodes from their active neighbors list. In the second

---

Algorithm 9.2 *Rank_Vcol*

1: **set of int** *free_cols* ← {1, ..., d_i + 1}; *received, lost_neighs* ← ∅; *curr_neighs* ← $\Gamma(i)$
2: **message types** *round, color, discard*
3: **boolean** *colored, round_recvd, round_over* ← *false*,
4: {round k for all nodes}
5: **while** ¬*round_over* **do**
6:      **receive** *msg(j)*
7:          **case** *msg(j).type* **of**
8:              round(k):   **if** ¬*colored* **then**
9:                                    **if** $i > max\{curr\_neighs\}$ **then**
10:                                       c_i ← *min{free_cols}*
11:                                       **send** *color(c_i)* to *curr_neighs*
12:                                       *colored* ← *true*
13:                                   **else send** *discard* to *curr_neighs*
14:                                   *round_rcvd* ← *true*
15:             color(c):   *received* ← *received* ∪ {j}
16:                              *free_cols* ← *free_cols* \ {c}
17:                              *lost_neighs* ← *lost_neighs* ∪ {j}
18:             discard: *received* ← *received* ∪ {j}
19:
20:      **if** (*round_recvd*) ∧ (*received* = *curr_neighs*) **then**
21:          *round_over* ← *true*; *curr_neighs* ← *curr_neighs* \ *lost_neighs*
22:          *round_rcvd* ← *false*; *received, lost_neighs* ← ∅
23:      **end if**
24: **end while**

---

round shown in (b), nodes 6, 5, and 2 are now the highest identifier nodes, and they select the minimum unused color c_2 as c_1 is used by their neighbors from the previous round. This procedure is repeated for rounds 3 and 4 as shown in (c) and (d), and the final coloring consists of four colors.

9.3.1.1 Analysis

Theorem 9.2 *The Rank_Vcol algorithm provides a legal coloring of an arbitrary graph with $O(\Delta + 1)$ colors in $O(n)$ rounds using $O(nm)$ messages.*

Proof We need to show that in each round of the algorithm, any highest identifier node can find a free color that is unused by its neighbors. The proof is trivial as in the extreme case where all neighbors of the lowest identifier node that has a degree of Δ, the color $\Delta + 1$ is available to color itself. In the worst case, we would have a linear network with decreasing identifiers, and there would be a sequential execution starting from the highest identifier node, continuing with the next highest one. The total number of rounds would then be n, resulting in a time complexity of $O(n)$ rounds. As there will be a constant time of traversals by the *color* and *discard* messages in each round, the total number of messages will be $O(nm)$ in $O(n)$ rounds.   □

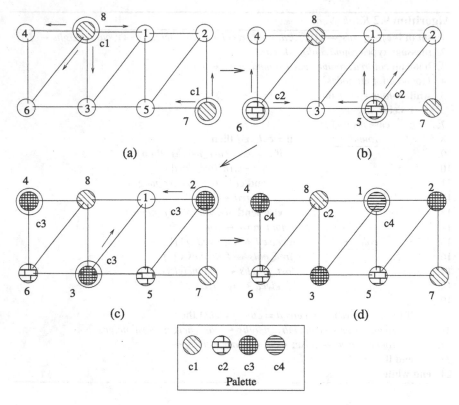

Fig. 9.2 A coloring example with *Rank_Vcol*

9.3.2 Random Vertex Coloring

The second synchronous algorithm, called *Rand_Vcol*, works in rounds, and we have an overlay spanning tree T as before. The general idea of this algorithm is that an uncolored node picks an unused color by their neighbors from their palette randomly and sends their choice to their neighbors. If there are no conflicts, the color for a node is decided, otherwise nodes stay in undecided state and repeat choosing a random color in the next round.

9.3.2.1 The First Version

In the first version of this algorithm, there are two phases within a synchronous round. In the first phase, all nodes exchange the *tentative(color)* messages to inform neighbors of the color they randomly select from available colors. When all these messages are received, a node can make a decision and decide on the color it has sent if there are no other neighbors that have selected the same color. Any node that has decided sends a *decide* message, and all others that have selected coinciding colors with neighbors send the *undecide* messages in the second phase of the round.

A detailed description of the algorithm is as follows. Initially, *free_cols* list at node i is initialized to a set of colors of size $d_i + 1$ in order where d_i is the degree of i and $d_i \leq \Delta$. In each round, if a node is not colored before, it picks an unused color c_i from the *free_cols* randomly and informs its neighbors of this choice by the *tentative* message. Each node waits *tentative* messages from all of its uncolored current neighbors (*curr_neighs*) and then checks if any other neighbor has picked the same color c_i. If not, it sends the *decide*(c_r) message to its uncolored neighbors to inform its choice. Otherwise, it sends the *undecide* message so that neighbors know when this communication phase is over. Different than most of the synchronous algorithms we have seen, there are two synchronization points in this round. The first is when node i receives a round message and all tentative messages from active neighbors in Line 13. At this point it can make a decision and send the *decide* or *undecide* message. When all these messages are collected at a node, *round_over* flag becomes true in Line 24, and the next round can be started. This process in each round continues for a number of rounds until each node is colored. A round of this algorithm, called *Rand_Vcol*, is shown in Algorithm 9.3.

9.3.2.2 The Second Version

In the first version of the algorithm, each active node sends two messages as *tentative* and *decide* or *undecide*. In an attempt to reduce the number of messages used by discarding the *undecide* message, the sending of the *decide* message can be delayed to the next round. In this second version, called *Rand2_Vcol*, each node that has not decided in a previous round sends the *tentative*(c_t) messages, and a node that has decided in a previous round shown by the *color_flag* sends the *decide*(c_d) messages to its active neighbors. However, there is the possibility that a decided color c_d of a node i may be the same as the tentative color c_t of a node j, and c_d is therefore included in the received colors list (*recvd_cols*) in this version. A round of this algorithm is shown in Algorithm 9.4.

This algorithm always computes a legal coloring in $O(\log n)$ rounds with probability approaching 1 as the number of vertices increases as shown in [9]. Luby [12] modified this algorithm so that at the start of each round every uncolored node is asleep and wakes up with probability $1/2$ and at the end of the round the uncolored nodes go back to sleep again. It was later shown that in each round of the algorithm, any uncolored vertex will be colored with the probability of at least $1/4$ [8].

9.3.3 A Simple Reduction Algorithm

In a network graph G where nodes have n unique identifiers, each identifier could represent the color of a node in the range $\{1, \ldots, n\}$, and hence a legal coloring of G

Algorithm 9.3 *Rand1_Vcol*

```
 1: set of int recvd_cols, lost_neighs, received ← ∅; curr_neighs ← Γ(i)
 2: message types round, tentative, decide, undecide
 3: boolean colored, round_recvd, round_over, phase1_end ← false
 4: {round k for all nodes}
 5: while ¬round_over do
 6:     receive msg(j)
 7:         case msg(j).type of
 8:             round(k):        if ¬colored then
 9:                                  pick cᵢ ∈ free_cols randomly
10:                                  send tentative(cᵢ) to curr_neighs
11:                                  round_recvd ← true
12:             tentative(c):    recvd_cols ← recvd_cols ∪ {c}
13:                              if round_recvd ∧ (recvd_cols = curr_neighs) then
14:                                  if ∄c ∈ recvd_cols where c = cᵢ then
15:                                      send decide(cᵢ) to curr_neighs
16:                                      colored ← true
17:                                  else send undecide to curr_neighs
18:                                  phase1_end ← true
19:             decide(c):       received ← received ∪ {j}
20:                              free_cols ← free_cols \ {c}
21:                              lost_neighs ← lost_neighs ∪ {j}
22:             undecide(c):     received ← received ∪ {j}
23:
24:     if (phase1_end) ∧ (received = curr_neighs) then
25:         round_over ← true; curr_neighs ← curr_neighs \ lost_neighs
26:         round_recvd ← false; received, recvd_cols, lost_neighs ← ∅
27:     end if
28: end while
```

---

is already achieved. However, by Brooke's theorem, we know that the coloring of G can be achieved using $O(\Delta + 1)$ colors. In this section, we provide an algorithm that reduces the coloring of a colored graph with colors same as identifiers of nodes to $O(\Delta + 1)$ colors. This algorithm, called *Redun_Vcol*, works in synchronous rounds numbered from n down to $\Delta + 1$. Initially, the color of a neighbor node that has an identifier in the range $\{1, \ldots, \Delta + 1\}$ is excluded from the free colors as these nodes will not change their colors, therefore, their colors may not be used. In each round, a node that has an identifier that equals the round number selects an unused minimum color c from free colors $1, \ldots, \Delta + 1$ and informs its neighbors by the message $color(c)$ that it has done so to prevent the selection of the same color by their neighbors in the next rounds as shown in Algorithm 9.5.

A node may have the following modes of execution in each round:

1. Its identifier is the same as the round number, in which case it reduces its color and informs all neighbors.
2. One of its neighbors' identifier is the same as the round number, in which case it waits for the *color* message from this neighbor node.

Algorithm 9.4 *Rand2_Vcol*

1: **set of int** *recvd_cols, received, lost_neighs* ← ∅; *curr_neighs* ← Γ(*i*)
2: **message types** *round, tentative, decide*
3: **boolean** *colored, color_flag, round_recvd, round_over* ← *false*
4: {round *k* for all nodes}
5: **while** ¬*round_over* **do**
6:     **receive** *msg(j)*
7:         **case** *msg(j).type* **of**
8:             <u>*round(k)*:</u>   **if** ¬*colored* **then**
9:                                 **if** *color_flag* **then**
10:                                    **send** *decide(c_i)* to *curr_neighs*
11:                                    *colored* ← *true*
12:                                **else pick** *c_i* ∈ *free_cols* randomly
13:                                    **send** *tentative(c_i)* to *curr_neighs*
14:                             *round_recvd* ← *true*
15:             <u>*tentative(c)*:</u>   *recvd_cols* ← *recvd_cols* ∪ {*c*}
16:                                 *received* ← *received* ∪ {*j*}
17:             <u>*decide(c)*:</u>   *received* ← *received* ∪ {*j*}
18:                                 *recvd_cols* ← *recvd_cols* ∪ {*c*}
19:                                 *free_cols* ← *free_cols* \ {*c*}
20:                                 *lost_neighs* ← *lost_neighs* ∪ {*j*}
21:
22:         **if** *round_recvd* ∧ (*received* = *curr_neighs*) **then**
23:             **if** ∄*c* ∈ *recvd_cols* where *c* = *c_i* **then**
24:                 *color_flag* ← *true*
25:             **end if**
26:             *round_over* ← *true*; *curr_neighs* ← *curr_neighs* \ *lost_neighs*
27:             *round_recvd* ← *false*; *received, recvd_cols, lost_neighs* ← ∅
28:         **end if**
29: **end while**

---

3. Neither its or one of its neighbor's identifiers is the same as the round number, in
   which case it does not do anything.

Also, it is possible that the *color* message may be received before the *round*
message. Algorithm 9.5 provides necessary controls so that all the above-described
execution modes are considered with uncertain message arrivals. Node *i* can predict
that it will receive the *color* message in a round as it knows the identifiers of all its
neighbors as we always assume. In order to make use of this condition, if a node
identifier *i* is not equal to the round number *col* but node *i* has a neighbor that has
an identifier *col*, *color_recvd* flag is made *false* to indicate that node *i* should wait
for the color message from its neighbor so that the sent color may be excluded from
the *free_cols* list.

Figure 9.3 displays the execution of Algorithm 9.5 in a network of eight nodes.
Initially, each node is colored by its unique identifier. In the first round of the algorithm, the round number is 8, node 8 recolors itself with the minimum unused color
from {1, ..., Δ + 1}, which is 2 and informs its neighbors of this choice to exclude 2

Algorithm 9.5 *Redun_Vcol*

1: **set of int** *free_cols* ← {1, . . . , $\Delta + 1$}; *curr_neighs* ← $\Gamma(i)$; *lost_neighs* ← ∅
2: **message types** *round, color*
3: **int** *c, col*
4: **boolean** *color_recvd* ← *true, round_recvd, round_over, colored* ←*false*
5: **for** *c* = 1 **to** $\Delta + 1$ **do**   ▷ exclude the neighbor colors of 1, . . . , $\Delta + 1$
6:      **if** $\exists j \in \Gamma(i)$ where *j = c* **then**
7:          *free_cols* ← *free_cols* \ {*c*}
8:      **end if**
9: **end for**
10: **for** *col = n* **down to** $\Delta + 1$ **do**
11:     {a single round for all nodes}
12:     **while** ¬*round_over* **do**
13:         **receive** *msg(j)*
14:             **case** *msg(j).type* **of**
15:                 <u>*round(col)*:</u>   **if** *i = col* **then**
16:                                     c_i ← *min*{*free_cols*}
17:                                     **send** *color(c_i)* **to** *curr_neighs*
18:                                 **else if** $\exists j \in curr\_neighs$ where *j = col* **then**
19:                                     *color_recvd* ← *false*   ▷ wait for *color* message
20:                                 *round_recvd* ← *true*
21:                 <u>*color(c)*:</u>   *free_cols* ← *free_cols* \ {*c*}
22:                                 *lost_neighs* ← *lost_neighs* ∪ {*j*}
23:                                 *color_recvd* ← *true*
24:             **if** *round_recvd* ∧ *color_recvd* **then**
25:                 *round_over* ← *true*; *curr_neighs* ← *curr_neighs* \ *lost_neighs*
26:                 *round_recvd* ← *false*; *color_recvd* ← *true*; *lost_neighs* ← ∅
27:             **end if**
28:     **end while**
29: **end for**

---

from their free color list. Similarly, nodes 7, 6, and 5 recolor themselves with colors 1, 3, and 4 as shown, and finally a legal coloring in the required range is obtained in (d).

9.3.3.1 Analysis

Theorem 9.3 *Redun_Vcol algorithm provides a legal coloring of an arbitrary graph with* $O(\Delta + 1)$ *colors in* $(n - \Delta - 1)$ *rounds using* $O(m)$ *messages.*

Proof Even if the neighbors of a node are colored from the set 1, . . . , Δ, there will still be a color available from this set. The number of rounds is equal to $(n - \Delta - 1)$ as the count of the *for* loop, and as each edge *e* will be traversed at most once by the *color* message, the message complexity is $O(m)$.   □

The algorithm *Redun_Vcol* is slow as it has a single point of activity at a time since the identifiers of the nodes are unique. However, it has a favorable message

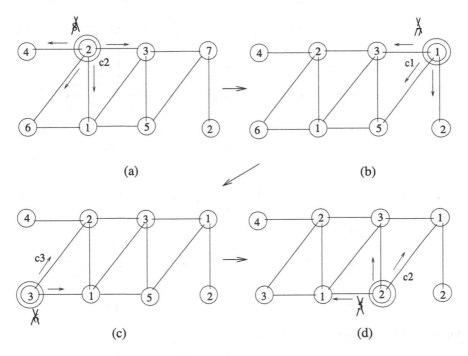

Fig. 9.3 A coloring example with *Redun_Vcol*

complexity as each edge is traversed at most once by the *color* message if it is incident to node i such that $\Delta + 1 \leq i \leq n$.

The same idea can be used to reduce coloring of a k-colored graph to $\Delta + 1$ colors. The structure of this algorithm (*Reduk_Vcol*) is similar to *Redun_Vcol*, and we will only show its operation in a sample graph. As can be seen in Fig. 9.4, the parallel execution of the nodes with the same color is possible in this case since non-adjacent nodes may have same colors. In this figure, nodes in a network of 8 nodes have labels showing their colors and each node is colored by $k = 7$ colors initially. The nodes with colors 7, 6, and 5 are recolored in the corresponding rounds, resulting in the graph of Fig. 9.4(d) with colors from $\{1, \ldots, \Delta + 1\}$. The time complexity of this algorithm is $k - \Delta - 1$ and requires $O(m)$ messages as before.

9.4 Edge Coloring

Edge coloring of a graph G requires each edge incident to the same vertex to be colored with different colors. We will attempt to have an identifier-based algorithm, called *Rank_Ecol*, to color edges of G. The key idea in this algorithm is to have the larger identifier at the end of an edge decide its color, and in each round, the highest identifier node in a neighborhood decides to color edges incident to it. The first thing to notice is that this highest identifier node may color more than one edge at

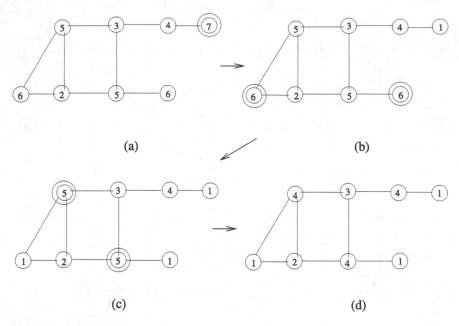

Fig. 9.4 A coloring example with *Reduk_Vcol*

once. Secondly, two nodes u and v may request the same color for edges $\{u, w\}$ and $\{v, w\}$ incident to the node w that would violate edge-coloring. In the algorithm we propose, if two or more higher identifier nodes request the same color for different edges incident, the lower identifier node accepts the highest identifier node offer and rejects the others.

The types of messages are *propose, unpropose, ack, reject, neigh_ack*. All other nodes that do not have all edges colored send the *unpropose* messages for synchronization. At this point, node u should wait for *ack* messages if the color was accepted by the node at the other end of the edge or *reject* otherwise. Instead of breaking a round into two phases, we will use a different strategy by odd and even numbered rounds where the *propose* and *unpropose* messages are sent and received in odd-numbered rounds; and the *ack* and *reject* messages are replied to the proposers in even-numbered rounds as shown in Algorithm 9.6 and Algorithm 9.7. Key to the operation of this algorithm is that each node should be aware of the colors of edges incident to all of its neighbors it has so that it will not propose the same colors in future rounds. In order to accomplish this, node i that accepts a proposal made by node j together with node i informs all of its active neighbors. The following describes the actions in an odd-numbered round of the algorithm.

1. At the reception of the *round* message, each node checks whether all edges incident to it are colored. If there are still uncolored edges, node i that finds it has the highest identifier among active neighbors selects tentative available colors from its free colors list (*free_cols*) and sends a different color to each active neighbor by the *propose(c)* message, where c is the tentative color.

Algorithm 9.6 *Rank_Ecol*: an odd numbered round

```
 1: int i, j, c                                    ▷ i is this node, j is the sender of a message
 2: set of int free_cols[Δ] ← {1, 2, ..., d_i} for each neighbor; received, recvd_cols,
    wait_acks, tent_cols[Δ] ← ∅; uncovd_edges ← all incident edges; lost_neighs ← ∅
 3: message types round, propose, unpropose, ack, reject, neigh_ack
 4: boolean round_over, round_recvd, prop_flag, reply_flag ← false
 5: for k = 1 to Δ do
 6: {rounds k, k + 1 for all nodes}
 7:   if (k mod 2) = 1 then
 8: while ¬round_over do
 9:      receive msg(j)
10:      case msg(j).type of
11:        round(k):    if uncovd_edges ≠ ∅ then
12:                       if i > max{curr_neighs} then              ▷ if i am largest
13:                         prop_flag ← true
14:                         ∀j ∈ curr_neighs                        ▷ send proposal to all
15:                           tent_cols[j] ← free_cols[j]
16:                         select c ∈ tent_cols[j]; send propose(c) to j
17:                         wait_acks ← wait_acks ∪ {j}
18:                         tent_cols[j] ← tent_cols[j] \ {c}
19:                       else send unpropose to curr_neighs
20:                     round_recvd ← true
21:        propose(c):  received ← received ∪ {j}
22:                     recvd_cols ← recvd_cols ∪ {⟨j, c⟩}
23:                     reply_flag ← true
24:        unpropose:   received ← received ∪ {j}
25:      if round_recvd ∧ (received = curr_neighs) then
26:         round_recvd ← false; round_over ← true; received ← ∅
27:      end if
28: end while
```

2. All nodes other than the highest identifier node send the *unpropose* messages to their active neighbors. The sum of all messages received as *propose* and *unpropose* should be equal to the number of active neighbors for this round to finish.

At the end of this round, each node may be classified as a *proposer*, *proposed*, or *other* node. In an even round, each type of node behaves differently.

1. *The proposed node*: Node i first checks the list *recvd_prop*. If any colors proposed already exist on its already colored edges, it sends a reject message to the proposer. It then checks if two or more proposers required the same color. In this case, it accepts the highest identifier node and sends the *reject* message to all others. For all remaining proposers in the *recvd_prop* list, it sends the *ack* message to confirm coloring.

2. *The proposing node*: Node i that receives an acknowledgement from node j to its proposal made in the previous round assigns the color c proposed to its edge $\{i, j\}$. It may also receive the *reject* messages from the proposed nodes. After col-

---

Algorithm 9.7 *Rank_Ecol*: an even-numbered round

1: **else**
2: **while** ¬*round_over* **do**
3:     **receive** *msg*(*j*)
4:     **case** *msg*(*j*).*type* **of**
5:         <u>*round*(*k*)</u>:   **if** *reply_flag* **then**   ▷ a proposed node
6:                         **while** ∃{⟨*s*, *c*⟩} ∈ *recvd_cols*|*c* ∉ *free_cols*
7:                             **send** *reject* to *s*; *recvd_cols* ← *recvd_cols* \ {*s*, *c*}
8:                         **while** ∃{⟨*s*, *c*⟩, ⟨*t*, *c*⟩} ∈ *recvd_cols*
9:                             *u* ← max{*s*, *t*}; *v* ← min{*s*, *t*}
10:                             *free_cols*[*u*] ← *free_cols*[*u*] \ {*c*}
11:                             *uncovd_edges* ← *uncovd_edges* \ {*i*, *u*}
12:                             *curr_neighs* ← *curr_neighs* \ {*u*}
13:                             **send** *ack*(*c*) to *u*; **send** *reject*(*c*) to *v*
14:                             *replied* ← *replied* ∪ {*u*, *v*}
15:                             *recvd_cols* ← *recvd_cols* \ {{*u*, *c*} ∪ {*v*, *c*}}
16:                         ∀(⟨*u*, *c*⟩) ∈ *recvd_cols*
17:                             **send** *ack*(*c*) to *u*
18:                             *free_cols* ← *free_cols* \ {*c*}
19:                             *replied* ← *replied* ∪ {*u*}
20:                         **send** *inactive* to *curr_neighs* \ *replied*
21:                         **else if** ¬*prop_flag*
22:                             **send** *inactive* to *curr_neighs*
23:                         *round_recvd* ← *true*
24:         <u>*ack*(*c*)</u>:   *free_cols*[*j*] ← *free_cols*[*j*] \ {*c*}   ▷ a proposer node
25:                         *acked_cols* ← *acked_cols* ∪ {*j*}
26:                         *uncovd_edges* ← *uncovd_edges* \ {*i*, *j*}
27:                         *curr_neighs* ← *curr_neighs* \ {*j*}
28:                         *wait_acks* ← *wait_acks* \ {*j*}
29:                         **if** *wait_acks* = ∅ **then**
30:                             **send** *neigh_ack*(*acked_cols*) to *curr_neighs*
31:         <u>*reject*(*c*)</u>:   *wait_acks* ← *wait_acks* \ {*j*}
32:                         **if** *wait_acks* = ∅ **then**
33:                             **send** *neigh_ack*(*acked_cols*) to *curr_neighs*
34:         <u>*neigh_ack*(*acols*)</u>: *free_cols*[*j*] ← *free_cols*[*j*] \ *acked_cols*   ▷ a neighbor node
35:                         *lost_neighs* ← *lost_neighs* ∪ *acols*
36:         <u>*inactive*</u>:
37:     **if** *msg*(*j*).*type* ≠ *round* **then**
38:         *received* ← *received* ∪ {*j*}
39:     **end if**
40:     **if** *round_recvd* ∧ *received* = *curr_neighs* **then**
41:         *acked*, *rejected*, *wait_acks*, *recvd_cols* ← ∅
            curr_neighs ← *curr_neighs* \ *lost_neighs*
42:         *prop_flag*, *reply_flag*, *round_recvd* ← *false*; *round_over* ← *true*
43:     **end if**
44: **end while**
45: **end if**
46: **end for**

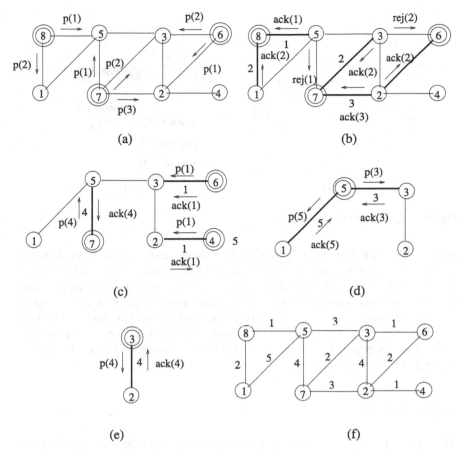

Fig. 9.5 *Rank_Ecol* execution example

lecting all replies, it informs its active neighbors by the *neigh_ack(acked_cols)*, where *acked_cols* is the list of accepted colors. It also informs the waiting neighbors by the *neigh_rej(n_rej)* message, where *n_rej* is the total number of *reject* messages received.

3. *Other nodes*: The other nodes that were waiting neighbors that had proposed in the previous round update their available colors lists when they receive the *neigh_ack(acked_cols)* messages for each proposer and synchronize by the reception of the *neigh_rej(n_rej)* messages. They send *inactive* messages to all neighbors for synchronization.

At the end of the algorithm, nodes i and j incident to an edge $\{i, j\}$ that is matched have their variable matched assigned true value as shown in Algorithm 9.6.

Algorithms 9.6 and 9.7 show the executions of an odd-numbered round and an even-numbered round consecutively, to edge-color a graph. Figure 9.5 displays the execution of *Rank_Ecol* over a sample graph with nodes $1, \ldots, 8$. In the first round,

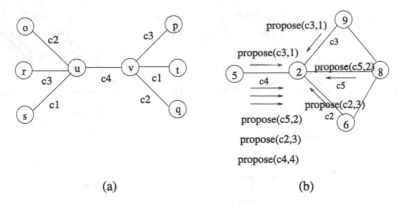

(a)   (b)

Fig. 9.6 Edge coloring with *Rank_Ecol*

nodes 8, 7, and 6 have the largest identifiers among their neighbors as shown by double circles and they are activated. Node 8 selects 1 and 2, node 7 selects 1, 2, 3, 4, and node 6 selects 1, 2, 3 as colors and send proposed messages to their neighbors in this round as shown in (a). In the second round, node 5 rejects the equal proposal by node 7 as node 8 has a greater identifier and sends *reject*(1) to node 7. Similarly, node 3 rejects the proposal by node 6 as node 7 has made a proposal for the same color. The new edges shown bold are added to the coloring, and finally the 5-edge-colored graph in (f) is obtained.

9.4.1 Analysis

Theorem 9.4 *The algorithm Rank_Ecol provides a legal edge coloring of a graph G using $O(\Delta)$ colors.*

Proof We need to show that in each round, any higher identifier node that proposes an uncolored edge to a lower identifier node has an available free color from a palette of (Δ) colors for a legal edge coloring. An edge e can be adjacent to $2\Delta - 2$ edges at most; however, $\Delta - 1$ of these edges are not adjacent and can use the same set of colors, which means that another color is needed to color the edge. Therefore, Δ colors are required to provide a legal coloring of an arbitrary graph with *Rank_Ecol*.   □

Figure 9.6(a) displays the situation where two nodes u and v of Δ degrees are adjacent and all their $2\Delta - 2$ neighbors are colored with two sets of colors as c_1, c_2, c_3, and c_4 is chosen to color the edge $\{u, v\}$. The total number of colors used is 4 as Δ.

Theorem 9.5 *The algorithm Rank_Ecol edge colors a graph G in $O(\Delta)$ rounds.*

Proof In the worst case, all of the proposals made by a higher identifier node u to a lower identifier node v may be rejected $\Delta - 1$ times as v may have $\Delta - 1$ neighbors that have higher identifiers than u, which may propose the same colors as u does. In this case, the proposal made at round Δ will have to be accepted as it will not violate any coloring that the neighbors of v have. Therefore, each edge e_{uv} may be colored in $O(\Delta)$ rounds.   □

Figure 9.6(b) shows the situation where nodes 5 and 2 are neighbors, and although node 2 is smaller than node 5, all its neighbors, which are nodes 6, 9, and 8, have larger identifiers than 5, and as shown in this example, if their proposals of color and the round number as *propose*(*color, round*) coincide with the proposals of node 5, the coloring of edge {5, 2} is delayed to the fourth round as Δ is 4.

Theorem 9.6 *The algorithm Rank_Ecol requires* $O(\Delta m)$ *messages to edge color a graph G.*

Proof In an odd-numbered round, there will be at most two messages over an edge as *propose* and *unpropose*. In an even-numbered round, there will again be at most two messages as *ack/reject* and *neigh_ack*, resulting in a total of $2\Delta m$ messages at most.   □

9.4.2 The Second Version

Grable et al. provided a synchronous, randomized distributed edge coloring algorithm [6]. In each round of this algorithm the following are performed:

- Each uncolored edge {u, v} randomly picks an unused colored c (a color that is not already incident to nodes u and v).
- If there are no collisions with other randomly selected colors for nodes u and v in this round, color c is assigned to edge {u, v}.
- Available colors for any uncolored edges incident to nodes {u, v} are updated by eliminating c from this list.

The algorithm continues until all edges are colored. In the implementation we propose, only the higher identifier node incident to an edge {u, v} picks a color c randomly. It then proposes c to the lower identifier node at the other end of edge {u, v}. A highest identifier node in a neighborhood can therefore send the *propose* messages to all its neighbors as in the algorithm *Rank_Ecol*; however, this algorithm allows the neighbors of the highest identifier nodes to propose to their lower neighbors; hence parallelism is enhanced, and there is parallelism even in the case of linear network with descending identifiers. A node that receives proposals of the same colors accepts the offer with the highest identifier node as before. The structure of this algorithm would be similar to the structure of *Rank_Ecol*, so we will only show its operation in a sample graph in Fig. 9.7. There are eight nodes, and each proposing node is shown by double circles. In round 1 as shown in (a), all nodes except nodes 4 and 1 are proposers. Out of these, all of them except nodes 7 and 8

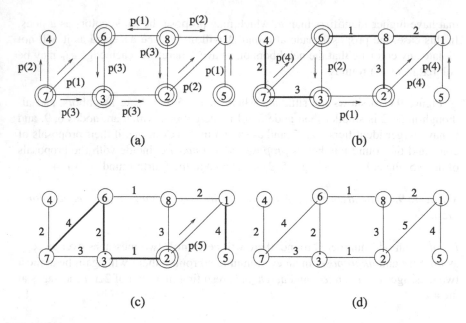

Fig. 9.7 Edge coloring algorithm second version example

also receive proposals from neighbors. Clearly, this implementation provides more parallelism. In four rounds, the final edge colored graph shown in (d) is obtained.

It was shown in [6] that this algorithm, called **S**, guarantees that the colors used for edge coloring is close to optimal value if the maximum degree of the graph is not too small. They also showed that with a minor modification of **S**, called **M**, where whenever a node runs out of colors, it can add a fresh new color to the available colors list. The algorithm **S** uses $(1 + \varepsilon)\Delta$ colors, where ε is a given positive constant, and finishes coloring of all edges in $O(\log n)$ rounds. Algorithm **M** can color edges of a graph in $O(\log \log n)$ rounds. Marathe et al. [13] conducted an experimental analysis of both **S** and **M** algorithms and concluded that they are both very fast.

9.5 Coloring Trees

We will now describe algorithms to color trees. The first algorithm colors a tree with two colors asynchronously. We then show two algorithms that are faster than this algorithm but color the trees with more colors.

9.5.1 A Simple Tree Algorithm

Clearly, two colors suffice to color a tree as shown in Fig 9.8. Algorithm 9.8 shows how a tree can be colored starting by the root coloring itself color c_0 and its chil-

Algorithm 9.8 *VcolTree_ASI*

```
 1: int i, j, parent, my_color
 2: set of int childs
 3: message types color
 4:
 5: if i = root then
 6:     send color(0) to childs
 7:     c_i ← 0
 8: else
 9:     receive color(c) from parent
10:     c_i ← 1 − c
11:     if childs ≠ ∅ then
12:         send color(c_i) to childs
13:     end if
14: end if
```

Fig. 9.8 Tree coloring with two colors

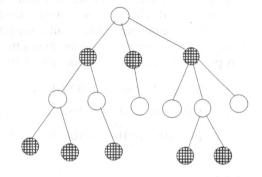

dren c_1 and grandchildren c_0, and so on. The time complexity of this algorithm is the depth of the tree, and in the worst case this would be $O(n)$ as the longest path between the root and a leaf would have $n - 1$ links in the case of a linear network. The message complexity is also similar as $n - 1$ edges of the tree will all be traversed once by the *color* messages. The problem with this algorithm is that the only concurrency is within the children of the same parent.

9.5.2 Six Coloring Algorithm

The *Six Coloring Algorithm* (*Six_Vcol*) is concurrently executed by all nodes of a tree T in synchronous rounds. Initially, all nodes have a color represented by their identifiers for a total of n colors. The algorithm starts by the root coloring itself with 0 and sending this color to its children. In every round thereafter, each node receives a color c_p from its parent and finds the smallest index k that its current color c_i differs from c_p. It then assigns a new color to itself by a bit string representing k concatenated with the value of c_i in the kth position. This new value of c_i is sent

---

Algorithm 9.9 *Six_Vcol*

1: **int** i, j, *parent*, *my_color*
2: **set of int** *childs*
3: **message types** *color*

4: **if** $i = root$ **then**   ▷ *root* assigns color 0
5:     $c_i \leftarrow 1$
6: **end if**
7: **while** $c_i \notin \{1, \ldots, 6\}$ **do**   ▷ do concurrently
8:     **receive** c_{parent}
9:     **find** the smallest index k that c_i and c_{parent} differ
10:     $c_i \leftarrow k$ in bits concatenated by the bit $c_i(k)$
11:     **send** c_i to children
12: **end while**

---

to children to be received in the next round and used for the same computation in the next round. This process continues until each node has a color in the range $\{0, \ldots, 5\}$, hence this algorithm is called the *Six Coloring Algorithm*. Algorithm 9.9 displays the operation of the algorithm in a tree T.

A possible reduction for the nodes with identifiers is shown below:

$$0001\ 0101\ 1001 \;\rightarrow\; 100001$$

$$1101\ 0101\ 1001 \;\rightarrow\; 110011 \;\rightarrow\; 0111 \;\rightarrow\; 01 \ \{k = 12, 4, \text{ and } 3\}$$

$$1010\ 1001\ 1001 \;\rightarrow\; 010011 \;\rightarrow\; 1000 \;\rightarrow\; 00 \ \{k = 12, 4, \text{ and } 3\}$$

9.5.2.1 Analysis

We analyze this algorithm as in [15].

Lemma 9.1 *The algorithm Six_Vcol provides a legal coloring in each round.*

Proof For a parent u and child v on the tree, let k_1 and k_2 be the two bit positions they differ from their parents. If $k_1 \neq k_2$, then u and v select colors that are different in their first components. Otherwise, if $k_1 = k_2$, then they differ in the second component.   □

The algorithm *Six_Vcol* provides a final coloring $c_i \in \{1, \ldots, 6\}$ for each node $i \in T$ with the final coloring consisting of six colors as shown in [15].

9.5.3 Six-to-Two Coloring Algorithm

In order to decrease the number of colors used in *Six_Vcol*, the algorithm *SixTwo_Vcol* may be used, which also works in concurrent synchronous rounds.

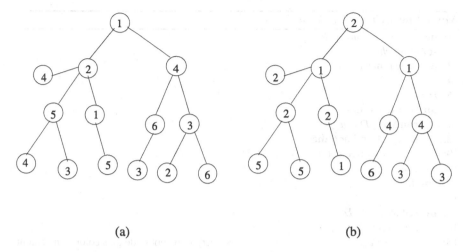

(a)   (b)

Fig. 9.9 An example of *Shift_Down* procedure

We first show the procedure *Shift_Down*, which for every nonroot node, colors the node with the color of its parent, the root then chooses a new color as shown in Fig. 9.10, which displays the operation of *Shift_Down* in a tree T, where the original colored tree with six colors by the *Six_Vcol* is shown in (a), and T after the shift down is shown in (b). As can be seen, the color values of the leaf nodes are discarded after the shift down.

The pseudocode for the procedure *Shift_Down* is shown in Algorithm 9.10. Initially, the *Six_Vcol* algorithm is executed on the tree T to provide a color $c_i \in \{1, \ldots, 6\}$ for each node i. Then, for each of the color values 4, 5, and 6, first, a shift down operation is performed, and then for each color value k, each node i checks whether its c_i value is equal to k. If this is true, the node i picks the first free color $c_i \in \{1, 2, 3\}$.

Figure 9.10 displays the execution of the *SixTwo_Vcol* algorithm in a sample tree, which is colored by six colors as shown in (a). After shift down operation, nodes with color 6 are colored with reduced colors as shown in (b) with double circles. Nodes that have color 5 are recolored in (c), and finally a single node with color 4 is recolored to result in the colored tree of (d).

9.5.3.1 Analysis

We analyze this algorithm as in [15].

Theorem 9.7 *The procedure Shift_Down does not disturb any legal coloring φ_m that exists of graph G.*

Proof An already legal coloring φ_m of G prior to *Shift_Down* means that any sibling of a parent has a different color than its parent. Assigning the color c_p of a parent to

---

Algorithm 9.10 *SixTwo_Vcol*

1: **int** i, j, *parent*, *my_color*
2: **set of int** *childs*
3: **message types** *color*
4:
5: **run** *Six_Vcol*
6: **for** $k = 4, 5, 6$ **do**
7:     **run** *Shift_Down*
8:     **if** $c_i = 4$ or 5 or 6 **then**
9:         $c_i \leftarrow$ a free color c from $\{1, 2, 3\}$
10:    **end if**
11: **end for**
12:
13: **procedure** *Shift_Down*
14:    **if** $i \neq root$ **then**
15:        $c_i \leftarrow c_p$ ▷ Every non-root node gets color of its parent
16:    **else**
17:        $c_i \leftarrow$ a free color c from $\{1, 2, 3\}$
18:    **end if**
19: **end procedure**

---

children would result in all children having a monochromatic color of c_p; therefore, as the children are not connected, legal coloring is obeyed for them. As between the parent and the children, the previous coloring φ_p was legal, the parent and children will also have a legal coloring with the new coloring φ_q. The only remaining issue is the case of the root, which now has the same color as its children as it does not have a parent and does not receive any colors. For this reason, after the nodes concurrently change their colors, the root chooses a new free color.   □

Theorem 9.8 *Each round of SixTwo_Vcol results in a legal coloring of the tree T.*

Proof The *Shift_Down* procedure does not disturb any legal coloring of T as shown by Theorem 9.7. In the color reduction phase, each node i that has a color from the set $\{4, 5, 6\}$ will find an available color c_i from $\{1, 2, 3\}$ as even if its parent and children are colored with two different colors, there will be a third available color. Also parallel recoloring of these nodes is legal as they are not adjacent.   □

The *SixTwo_CSI* algorithm provides a three-coloring of a tree T in $O(\log^* n)$ rounds as shown in [15].

9.6 Self-Stabilizing Vertex Coloring

In this section, we first describe a self-stabilizing algorithm to color a planar graph with six colors and then a second self-stabilizing algorithm to color arbitrary graphs with $O(\Delta + 1)$ colors.

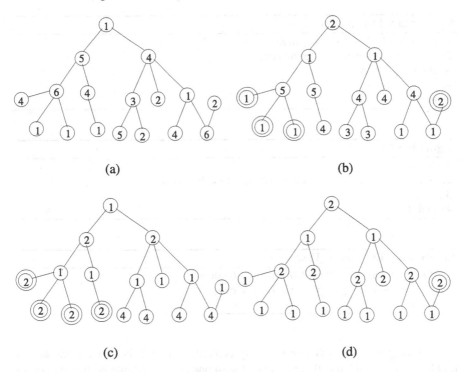

Fig. 9.10 Operation of *SixTwo_Vcol* algorithm

9.6.1 Coloring Planar Graphs

Karaata et al. [4] proposed a two-stage self-stabilizing algorithm that works under a central daemon to color undirected planar graphs. In the first stage, a directed acyclic graph is obtained from an undirected planar graph, and this graph is colored with six colors in the second stage. Each node i is assigned an arbitrary integer value x_i to determine the direction of its incident edges in the first stage. An edge e_{ij} between nodes i and j is assumed to be directed from the lower x-valued node to the higher-valued node, and ties are broken by the unique node identifier values. Starting from an arbitrary state, node i may have more than five outgoing edges, in which case it changes the direction of all its outgoing edges to inward directions by assigning a value greater than any of the values assigned to its outgoing edges as shown in Algorithm 9.11. The out_i variable shows the number of outgoing edges that node i has, and $xvals$ is the set of x values that the outgoing edges from node i has.

In the second stage of the algorithm (*SS1_Vcol*) shown in Algorithm 9.12, each node i checks whether it has the same color with any of its successor edges. If this is valid, it obtains the first color that is available from the set *colors* that is not used by any of its successor nodes (*succrs*). The *colors* set has values $\{1, \ldots, 6\}$ assuming that any graph can be colored by $\Delta + 1$ colors.

Algorithm 9.11 DAG_SS

1: $xvals \leftarrow \bigcup_x \forall x \in outgoing\_edges$
2: $out_i \leftarrow$ the number of outgoing edges
3: **if** $out_i \geq 5$ **then**
4:     $x_i \leftarrow (max\{xvals\} + 1)$
5: **end if**

Algorithm 9.12 $SS1\_Vcol$

1: $colors \leftarrow \{1, \ldots, 6\}$
2: **if** $\exists j \in succrs_i \mid (c_i = c_j) \wedge (k \in (colors \setminus succols_i))$ **then**
3:     $c_i \leftarrow k$
4: **end if**

Algorithm 9.13 $Grundy\_Vcol$

1: **if** $c_i \neq min\{col \geq 1 \mid (\forall j \in \Gamma(i))(c_j \neq col)\}$ **then**
2:     $c_i \leftarrow min\{col \geq 1 \mid (\forall j \in \Gamma(i))(c_j \neq col)\}$
3: **end if**

Assuming that graph G is not legally colored, there will be at least one node i that has $c_i = c_j$, and also there will be at least one color available in the set $colors \setminus succols_i$ since $succols = 5$ and $colors$ have six distinct colors. This implies that whenever a node has the same color as one of its successors, it will have an available legal color to color itself. Furthermore, if the colors of the successors of a node i do not change, i will reach a stable state by one move [4].

9.6.2 Coloring Arbitrary Graphs

In this section, we describe the first linear-time self-stabilizing vertex coloring algorithm proposed by Hedetniemi [7], which uses $\Delta + 1$ colors and stabilizes in a maximum of n moves. This algorithm is based on *Grundy Coloring*, which is defined as follows.

Definition 9.3 (Grundy Coloring) In *Grundy Coloring*, node i is colored with the minimum color that is not used by any of its neighbors. Formally, the color of node $c_i = min\{col \geq 1 \mid (\forall j \in \Gamma(i))(c_j \neq col)\}$.

In Algorithm 9.13 (*Grundy_Vcol*), each node applies the Grundy Coloring Rule until a stable condition is reached where a node that has a color different from the smallest integer not taken by any of its neighbors sets its color to that integer as described in [7].

Algorithm 9.14 *Hedet_Vcol*

1: **if** $c_i \in \{c_j | j \in \Gamma(i)\} \vee c_i > d_i + 1$ **then**
2:     **if** $\{c_j | j \in \Gamma(i)\} = \{1, \ldots, d_i\}$ **then**
3:         $c_i \leftarrow d_i + 1$
4:     **else set** $c_i \in \{1, \ldots, d_i\} - \{c_j | j \in \Gamma(i)\}$
5:     **end if**
6: **end if**

9.6.2.1 Analysis

A node would increase its color value only if all of its neighbors are colored with colors $\{1, \ldots, c_i\}$. A node can make a maximum of $d_i + 1$ decreasing moves. It can be shown that a node can in fact make a maximum of $d_i + 1$ moves.

Theorem 9.9 *Grundy_SS finds a legal coloring of a graph in* $O(2m + n)$ *moves.*

Proof Any move by node i results in a coloring $c_i \leq d_i + 1$, and node i can make $O(d_i + 1)$ consecutive moves, which all decrease the value of c_i. Since each node will make a maximum of $d_i + 1$ moves, the total number of moves will be $\sum_{i=1}^{n}(d_i + 1) = 2m + n$. A detailed proof as a consequence of lemmas can be found in [7]. $\square$

Based on the Grundy Coloring, Hedetniemi et al. proposed an anonymous algorithm (*Hedet_SS*) that works under a central daemon. Each node i in this algorithm checks whether it has a color equal to any of its neighbor colors. If this condition is met or a node has a color greater than $\Delta + 1$, it further checks whether the colors of neighbors have filled the colors $\{1, \ldots, d_i\}$. If this is true, it can select the color $d_i + 1$ as this is the only choice. Otherwise, if one or more neighbors do not have a color assigned in the range $\{1, \ldots, d_i\}$, node i picks the first available color not assigned to any of the neighbors as shown in Algorithm 9.14.

Figure 9.11 displays the operation of *Hedet_SS* in a network with six nodes, where labels represent colors, and double circles show the privileged nodes at each step. One of the nodes has a degree d_i of 2 and a color of 2, which its neighbor has, and it is enabled by the central daemon as shown by the double circle in (a). It changes its color to the color $d_i + 1$, which is 3 since all its neighbors have used the colors $\{1, \ldots, d_i\}$, which are 1 and 2, as displayed in (b). The other privileged node has a color of 5, which is greater than its $d_i + 1$, which is 4. It is enabled and chooses a color of $d_i + 1$, which is 4 as shown in (c). The lastly enabled node 6 can color itself with 1 as its neighbors have not used it, and the final $\Delta + 1$ colored stabilized graph, where there are no privileged nodes, is displayed in (d).

Theorem 9.10 *Hedet_SS finds a legal* $\Delta + 1$ *coloring of a graph in* $O(n)$ *moves.*

Proof When a node is privileged so that it can make a move, it will choose a color from $\{1, \ldots, d_i + 1\}$. Once it makes a move, it will be colored and stay in that state because coloring of a neighbor node does not affect its state. Therefore, as each

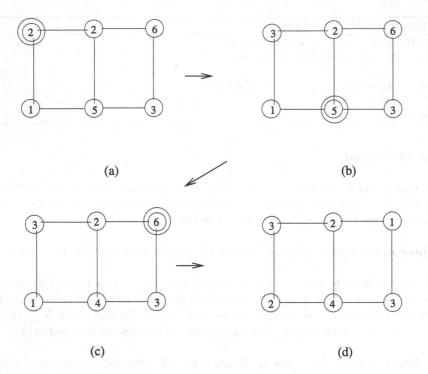

(a)

(b)

(c)

(d)

Fig. 9.11 Execution of *Hedet_SS*

node will make a maximum of one move, the total number of moves is $O(n)$. In stabilization, there would be at most $\Delta + 1$ colors since if the coloring uses more than $\Delta + 1$ colors, a node becomes privileged. A detailed proof as a consequence of lemmas can be found in [7].   □

9.7 Chapter Notes

We have seen rank-based, random, reduction-based, and self-stabilizing vertex coloring algorithms and an edge coloring algorithm in this chapter. The edge coloring algorithm is more complicated than others as coloring an edge is related to the neighbors of the neighbors of a node and any decision should be propagated.

Although it has been studied extensively using sequential algorithms, vertex coloring in distributed setting remains an active area of research due to a wide range of applications it has in computer networks. For deterministic vertex coloring, it was shown in [10] that $O(\Delta^2)$ coloring can be achieved in $O(\log n)$ rounds for bounded-degree graphs. In [14], a graph G is first partitioned into disjoint forests that are colored using three colors concurrently in the first phase. Using these colors, G is recolored with $O(\Delta + 1)$ colors in $O(\Delta^2 + \log^* n)$ rounds. Recently, algorithms to

Fig. 9.12 Example graph for Exercise 1

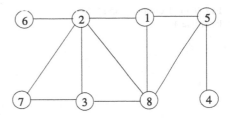

Fig. 9.13 Example graph for Exercise 2

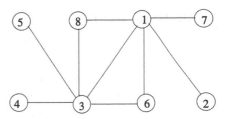

color graphs in using $O(\Delta + 1)$ colors in $O(\Delta + \log^* n)$ time were provided in [10] and [2] independently.

A distributed edge coloring algorithm that achieves $2\Delta - 1$ coloring in $O(\Delta + \log^* n)$ rounds is proposed as well as vertex coloring algorithm with $O(\Delta + 1)$ colors in $O(\Delta^2 + \log^* n)$ rounds in [14].

9.7.1 Exercises

1. Show the step-by-step execution of *Seq_Vcol* algorithm in the sample graph of Fig. 9.12.
2. Provide a pseudocode for a synchronous distributed algorithm that greedily selects nodes with the highest degrees first to color. Show the synchronization messages for this algorithm and work out the time and message complexities. Show also step-by-step execution of this algorithm in the example graph of Fig. 9.13.
3. Provide an FSM-based implementation of *Rand_Vcol* by drawing the FSM diagram and writing the pseudocode for this algorithm. Compare this pseudocode with the code of *Rand_Vcol*.
4. Provide a pseudocode for an algorithm that reduces colors of a k-colored graph to a maximum of $\Delta + 1$ colors such that all nodes send their colors and receive colors of neighbors in every round and then decide their color for the next round. Work out the time and message complexities of this algorithm and compare its efficiency with the *Rand_Vcol* algorithm. Show also the operation of this algorithm in the example graph of Fig. 9.14, which is colored by six colors as shown by the labels of the vertices.
5. Show the execution of *SixTwo_Vcol* algorithm step by step in the example tree of Fig. 9.15, which is colored by six colors as shown by the labels of the vertices.

Fig. 9.14 Example graph for
Exercise 4

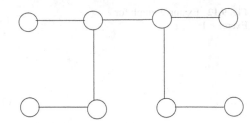

Fig. 9.15 Example graph for
Exercise 5

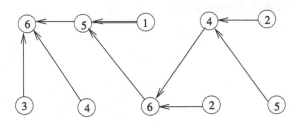

References

1. Appel K, Haken W (1977) Solution of the four color map problem. Sci Am 237(4):108–121
2. Barenboim L, Elkin M (2010) Deterministic distributed vertex coloring polylogarithmic time. In: Proc 29th annual ACM symposium on principles of distributed computing (PODC), Zurich, Switzerland. ACM, New York, pp 410–419
3. Bollobas B (1979) Graph theory. Springer, New York
4. Ghosh S, Karaata MH (1993) A self-stabilizing algorithm for coloring planar graphs. Distrib Comput 7:55–59
5. Goldberg AV, Plotkin SA, Shannon GE (1988) Parallel symmetrybreaking in sparse graphs. SIAM J Discrete Math 1(4):434–446
6. Grable D, Panconesi A (1997) Nearly optimal distributed edge-coloring in $O(\log^{\log n})$ rounds. Random Struct Algorithms 10(3):385–405
7. Hedetniemi ST, Jacobs DP, Srimani PK (2003) Linear time self-stabilizing colorings. Inf Process Lett 87:251–255
8. Johansson O (1999) Simple distributed $\Delta + 1$ coloring of graphs. Inf Process Lett 70(5):229–232
9. Karp RM (1991) Probabilistic recurrence relations. In: Proc 23rd annual ACM symposium on theory of computing (STOC 91), pp 190–197
10. Kuhn F (2009) Weak graph colorings. In: Proc 21st annual ACM symposium on parallelism in algorithms and architectures. ACM, New York, pp 138–144
11. Linial N (1992) Locality in distributed graph algorithms. SIAM J Comput 21(1):193–201
12. Luby M (1993) Removing randomness in parallel computation without a processor penalty. J Comput Syst Sci 47:250–286
13. Marathe V, Panconesi A, Risinger LD Jr (2000) An experimental study of a simple, distributed edge coloring algorithm. In: Proc SPAA 2000, pp 166–175
14. Panconesi A, Rizzi R (2001) Some simple distributed algorithms for sparse networks. Distrib Comput 14(2):97–100
15. Peleg D (2000) Distributed computing: a locality-sensitive approach. SIAM, Philadelphia. ISBN: 0-89871-464-8
16. Skiena SS (2008) The algorithm design manual, 2nd edn. Springer, Berlin. ISBN: 1-84800-069–978-1-84800-8. Chapter 16

Chapter 10
Maximal Independent Sets

Abstract An independent set of a graph is a subset of its vertices such that there are not any two adjacent vertices in this set. Finding the maximal independent set of a graph has many important applications such as clustering in wireless networks, and independent sets can also be used to build other graph structures. In this chapter, we describe rank-based, randomized, and self-stabilizing distributed algorithms to form maximal independent sets of graphs.

10.1 Introduction

An independent set can be formally defined as follows.

Definition 10.1 (Independent set) An *independent set* (*IS*) or a *stable set* of a graph $G(V, E)$ is a subset IS of the vertices of V such that there is no edge of G that joins any two vertices of S.

Definition 10.2 (Maximal and maximum independent sets) A *maximal* (MIS) of a graph G cannot be enlarged any further. The size of an independent set is the number of vertices it contains. A *maximum independent set* (*MaxIS*) is the largest independent set for a given graph G, and its size is denoted by $\alpha(G)$.

Finding the maximum independent set of a graph is an NP-hard optimization problem, and deciding whether a graph has a MIS of size k is an NP-complete problem. A set is independent if and only if it is a clique in the complement of the graph, and also a set is independent if and only if its complement is a vertex cover. The sum of $\alpha(G)$ and the size of minimum vertex cover ($\beta(G)$) is the number of vertices in the graph. Figure 10.1 displays IS examples where (a) is a MIS of size 1, (b) is a MIS of size 2, and (c) is a MaxIS of size 3.

In this chapter, we start by inspecting a simple sequential algorithm to find MIS which runs in polynomial time. We then describe a distributed MIS algorithm that uses identifiers of the nodes to find MIS. Randomization provides algorithms with better time complexities as shown by two synchronous randomized algorithms. The

K. Erciyes, *Distributed Graph Algorithms for Computer Networks*,
Computer Communications and Networks, DOI 10.1007/978-1-4471-5173-9_10,
© Springer-Verlag London 2013

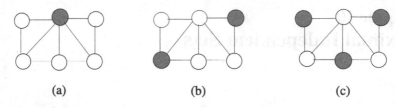

(a)   (b)   (c)

Fig. 10.1 Independent set examples

Algorithm 10.1 *Seq_MIS*

1: Input $G(V, E)$
2: $S \leftarrow V, MIS \leftarrow \varnothing$
3: **while** $S \neq \varnothing$ **do**
4:     **select** an arbitrary vertex $u \in S$
5:         $S \leftarrow S \setminus \{u \cup \Gamma(u)\}$
6:         $MIS \leftarrow MIS \cup \{u\}$
7: **end while**

---

first algorithm is based on random decisions by the nodes with the probability related to their degrees, and in the second randomized algorithm, nodes draw random values to be included in the MIS. All synchronous operations are carried over a spanning tree that is constructed prior to these algorithms, by any of the algorithms described in Chap. 4. We then provide three self-stabilizing algorithms in chronological order to build a MIS of a graph.

10.2 The Sequential Algorithm

The sequential algorithm is based on the observation that whenever a vertex u is included in the MIS, its neighbors should be excluded. The algorithm *Seq_MIS* shown in Algorithm 10.1 selects a vertex u arbitrarily and includes it in the MIS. Since this node is now in the MIS, any vertex in its neighborhood with its adjacent edges should be removed from graph G.

Figure 10.2 displays an example execution of *Seq_MIS* in an example graph of eight nodes. Node 4 is selected to be in MIS in the first iteration and is removed from the vertex set S with its neighbors 1 and 2. In the second iteration, node 6 is selected and is removed from S with neighbor 8. The third iteration picks node 3, which is removed from S with 5. The remaining node 7 is included in MIS finally, and the algorithm stops as $S = \varnothing$.

Theorem 10.1 *The time complexity of Seq_MIS is $O(n)$.*

Proof In the case of a linear network, there will be at most $\lceil n/2 \rceil$ selected nodes, and therefore, the time taken will be $O(n)$. Figure 10.3 displays such a situation

Fig. 10.2 *Seq_MIS*
execution example

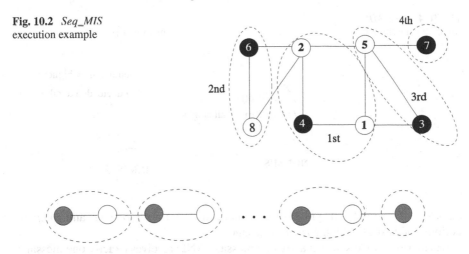

Fig. 10.3 Linear network independent set example

where dark nodes are the MIS nodes, and each iteration with the removed vertices
and edges from the graph are shown by dashed ellipses.   □

10.3 Rank-Based Distributed MIS Algorithm

In this section, we will describe a rank-based distributed algorithm to find MIS of a
network graph. The identifiers of the nodes are used to determine which node will
enter the MIS. Whenever a node enters the MIS, its neighbors should be excluded
from MIS and should not participate in the selection process in future rounds. These
two different phases within a round can be achieved by either deferral of the resigning of neighbors to the next round, having two distinct phases within a round
using flag variables or having odd- and even-numbered rounds where nodes may
enter MIS in odd rounds and neighbors resign in even rounds. Out of these three
methods, we will describe the first one as follows.

Our first distributed algorithm, implemented using an FSM and called *Rank_MIS*,
executes in synchronous rounds, selecting the highest identifier node among neighbors to be included in the MIS in each round. The algorithm runs until all nodes
determine whether they are in MIS or not. In each round, every active node *u* that
has not decided yet checks the states of its current active neighbors with higher
identifiers. If all of them have decided not to be in MIS, it decides to be in MIS and
notifies all its current neighbors of this decision. Each node can be in one of the
three states as undecided (UNDEC) initially, in INMIS State if it decides to be in
MIS, and in NONMIS state if it decides not to be in MIS as shown in Fig. 10.4.

Algorithm 10.2 shows a possible implementation of a single round of *Rank_MIS*
algorithm in an asynchronous setting. Four types of messages are *round, decide, un-*

Fig. 10.4 *Rank_MIS*
finite-state machine

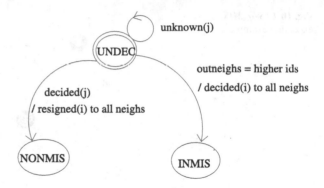

decide, and *resign*. Each synchronous round consists of *send*, *receive*, and *compute* actions and is started by the *round* message.

Each active node sends exactly one message AND receives exactly one message in each round. At the end of each round, any decided node should be removed from the graph, resulting in a smaller graph for the next round, and this is accomplished by deleting the nodes in MIS from the current neighbor list (*currneighs*). In each round, node i in UNDEC state that receives the *round* message checks whether its neighbors in NONMIS state are the same as its neighbors that have higher identifiers than itself. If this is true, meaning that all its higher neighbors have given up being in MIS due to a neighbor in MIS, node i decides to go into INMIS state and sends the *decide* message to all its active neighbors, otherwise, it sends the *undecide* message to its current neighbors. A node that has entered NONMIS state for the first time sends the *resign* message to all its *currneighs* in the next round to enable them to delete it from their *currneighs* and add to neighbors with higher identifiers that have entered *NONMIS* state. In order to synchronize the end of a round, all senders of the received messages are kept in the set *received*.

Figure 10.5 displays an example operation of Algorithm 10.2 in a sample network of eight nodes with identifiers $1, \ldots, 8$. It is assumed that a spanning tree T is initially constructed at root node 1 to send the synchronization messages such as *round* and *upcast* as shown in (a). Each node knows the identifiers of its neighbors and stores its higher identifier neighbors in *higher_ids*. Each node starts from the initial UNDEC state, and each message is tagged with the time frame (round) it occurs. The root node 1 starts the algorithm by sending the *round*(1) message to its only child, node 2. The following shows the actions in each round:

1. *Round 1*: Nodes 8 and 7 find that their higher identifier neighbors set is empty, send *decide* to their neighbors 2 and 6, which enter NONMIS state but defer sending the *resign* message till the next round. All other nodes send the *undecide* messages to their neighbors as their higher identifier neighbors are in UNDEC state.
2. *Round 2*: Nodes 2 and 6 send the *resign* messages to their active neighbors, which are nodes 3, 4, and 5. The first two rounds are shown in (b).
3. *Round 3*: Node 1 finds that its *higher_ids* list equals its *out_neighs* list (node 2) and enters MIS, but it does not send any *decide* message because its *curr_neighs*

Algorithm 10.2 *Rank_MIS*

 1: **states** *UNDEC, INMIS, NONMIS*
 2: **int** i, j, k, *currstate* $\leftarrow$ *UNDEC*
 3: **set of int** *out_neighs, lost_neighs, received, lost_neighs* $\leftarrow \varnothing$; *curr_neighs* $\leftarrow \Gamma(i)$;
    higher_ids $\leftarrow$ all higher id neighbors
 4: **message types** *round, decide, resign, undecide*
 5: **boolean** *outflag, round_recvd, round_over* $\leftarrow$ *false*
 6: {round k for all nodes}
 7: **while** $\neg$*round_over* **do**
 8:     **receive** *msg*(j)
 9:     **case** *currState* **of**
10:         *UNDEC*:
11:             **case** *msg*(j).*type* **of**
12:                 *round*(k):   **if** *higher_ids* $=$ *out_neighs* **then**
13:   *currstate* $\leftarrow$ *INMIS*
14:   **send** *decide*(k) to *curr_neighs*
15:                                     **else send** *undecide*(k) to *curr_neighs*
16:                                     *round_recvd* $\leftarrow$ *true*
17:                 *decide*(k):   *currstate* $\leftarrow$ *NONMIS, outflag* $\leftarrow$ *true*
18:                                     *lost_neighs* $\leftarrow$ *lost_neighs* $\cup \{j\}$
19:                                     *received* $\leftarrow$ *received* $\cup \{j\}$
20:                 *resign*(k):   *lost_neighs* $\leftarrow$ *lost_neighs* $\cup \{j\}$
21:                                     *received* $\leftarrow$ *received* $\cup \{j\}$
22:                                     **if** $j > i$ **then** *out_neighs* $\leftarrow$ *out_neighs* $\cup \{j\}$
23:                 *undecide*(k):   *received* $\leftarrow$ *received* $\cup \{j\}$
24:
25:         *NONMIS*:
26:                 *round*(k):   **if** *outflag* **then send** *resign*(k) to *currneighs*
27:   *outflag* $\leftarrow$ *false*
28:     **if** *currstate* $=$ *UNDEC* $\wedge$ *round_recvd* $\wedge$ (*received* $=$ *currneighs*) **then**
29:         *round_over* $\leftarrow$ *true*; *curr_neighs* $\leftarrow$ *curr_neighs* \ *lost_neighs*
30:         *round_recvd* $\leftarrow$ *false*; *received, lost_neighs* $\leftarrow \varnothing$
31:     **end if**
32: **end while**

---

    list is empty. Similarly, node 5 enters MIS and notifies its only current neighbor, node 4, with the *decide* message.

4. *Round 4*: Node 4 sends the *resign* message to its only active neighbor, node 3. The rounds 3 and 4 are shown in (c). In round 5, node 3 finds that it can enter MIS as its all higher identifier neighbors (nodes 4 and 6) have resigned, which completes the building of the MIS for this graph as shown in (d).

    It should be noted that in order to provide synchronization in each round, sending of the *resign* message is deferred to the next round. The execution of *Rank_MIS* has taken four rounds for this example. The time complexity can be as high as n rounds. The usage of the synchronization messages *upcast* and *finish* would be favorable, in which case the root node does not need to know the number of nodes in the network.

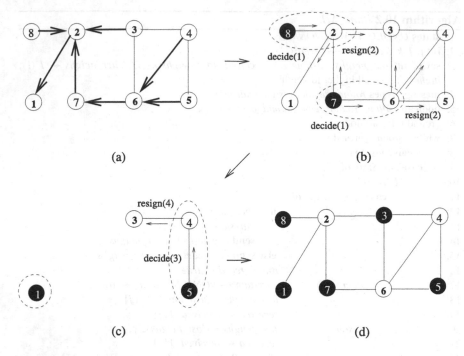

(a)   (b)

(c)   (d)

Fig. 10.5 *Rank_MIS* execution example

In this case, any node that enters the state *INMIS* or *NONMIS* would send the *finish* message to its parent, and when all the *finish* messages are convergecast to the root, it would terminate the algorithm.

10.3.1 Analysis

Theorem 10.2 *Algorithm Rank_MIS correctly constructs a MIS in* $O(n)$ *rounds, and its message complexity is* $O(nm)$.

Proof In the closed neighborhood of a node, there will be only one node joining MIS that will be the highest identifier node or the node with all of its higher identifier neighbors having given up entering the MIS. This ensures the correct operation of the algorithm as there will be no more than one node entering the MIS among the neighbors in any round.

We would have a sequential operation of the algorithm in the worst case where all nodes are ordered in a line in decreasing order of identifiers. The highest identifier node will first enter MIS, followed by the neighbor node resigning in the second round, followed by the third node entering the MIS in the third round, resulting in a total of $O(n)$ rounds as time complexity. Each edge of the network graph will be traversed at most once by a *decide* message and a constant number of *undecide* messages resulting in a message complexity of $O(nm)$ in total.   □

There is a minor issue with this algorithm, which is the deferral of the *resign* message to the next round. Any node i that finds that its neighbor j has decided in round k sends the *resign* message in the $(k + 1)$th round since each node waits exactly one message from its current neighbors in each round for proper synchronization. However, there may be a neighbor node t of node i that has the second highest identifier in its neighborhood after i, in which case it has to wait until next round. This case does not disrupt the operation of this algorithm but results in an extra round; however, the time complexity is still $O(n)$. In order to remedy this situation, we could have two phases in each round so that any node that resigns sends the *resign* message in the second phase, and nodes that do not resign send a new message *unresign* (see Exercise 1), or a new version of the algorithm may be designed as described next.

10.4 The First Random MIS Algorithm

The first randomized distributed algorithm we show is a simplified version of Luby's algorithm [7] called *Rand1_MIS*. The algorithm operates in synchronous rounds as described below:

1. Each node u marks itself with probability $1/(2\,dval(u))$, where $dval(u)$ is the current degree of u.
2. If there is not a higher-degree neighbor of u that is marked, node u joins the MIS. If a higher-degree neighbor of u is marked, node u unmarks itself, where ties are broken by identifiers if they have the same degree.
3. Any node that has joined the MIS and its neighbors are deleted from the active neighbor node lists since they cannot join the MIS anymore.

This choice favors nodes with lower degrees as choosing nodes with higher degrees would reduce the size of the MIS. However, once a higher-degree node is marked, it has a better chance to join MIS. The FSM of the algorithm is shown in Fig. 10.6, where each node starts from the unmarked (*UNMAK*) state and goes into MARKED state if it randomly has decided based on its current active degree. If all higher-degree neighbors are in *UNMAK* state, a marked node can enter INMIS state, and the state for a node that has decided not to join as a result of a neighbor that has joined is NONMIS.

A detailed operation of this algorithm for a single round is shown in Algorithm 10.3. The message *info* is used for a node to send its current degree and its current state to its current neighbors in each round. It is assumed that the start of each round is controlled by a single initiator by broadcasting the message *round* over the spanning tree T, and the end of a round is determined by the arrival of a converged *upcast* message to the initiator over T.

An example operation of *Rand1_MIS* is shown in Fig. 10.7 with eight nodes numbered $1, \ldots, 8$. Nodes 8, 3, 4, 2 mark themselves (shown by double circles) in

Algorithm 10.3 *Rand*1_*MIS*

1: **states** *MARKED, UNMAK, INMIS, NONMIS*; *currstate* ← *UNMAK*
2: **message types** *round, info, resign, decide, undecide*
3: **int** *i, j*; *currdeg* ← $|\Gamma(i)|$
4: **set of int** *recvd*1, *recvd*2, *degs*, *states* ← ∅; *currneighs* ← Γ_i
5: **boolean** *round_recvd, phase*1*over, round_over* ← *false*
6:
7: **while** ¬*round_over* **do**
8:    **receive** *msg*(*j*)
9:    **case** *currstate* **of**
10:
11:       *UNMAK*:
12:
13:           <u>*round*(*k*)</u>:   *currstate* ← *MARKED* w.p. 1/(2*currdeg*(*i*))
14:                            **send** *info*(*k, currdeg, currstate*) to *currneighs*
15:                            *round_recvd* ← *true*
16:
17:       *UNMAK/MARKED*:
18:           <u>*info*(*k, d, sta*)</u>:   *recvd*1 ← *recvd*1 ∪ {*j*}
19:                            *states* ← *states* ∪ {*sta*}, *degs* ← *degs* ∪ {*d*}
20:           <u>*resign*(*k*)</u>:   *lost_neighs* ← *lost_neighs* ∪ {*j*}
21:                            *currdeg* ← *currdeg* − 1
22:                            *recvd*1 ← *recvd*1 ∪ {*j*}
23:           <u>*decide*(*k*)</u>:   *currstate* ← *NONMIS, outflag* ← *true*
24:                            *recvd*2 ← *recvd*2 ∪ {*j*}
25:           <u>*undecide*(*k*)</u>: *recvd*2 ← *recvd*2 ∪ {*j*}
26:
27:       *NONMIS*:
28:           <u>*round*(*k*)</u>:   **if** *outflag* **then send** *resign*(*k*) to *currneighs*
29:                                      *outflag* ← *false*
30:
31:    **if** *round_recvd* ∧ (*recvd*1 = *currneighs*) **then**
32:      **if** *currstate* = *MARKED* **then**
33:        **if** ∀*x* ∈ *currneighs* where *states*(*x*) = *MARKED* ∧ *degs*(*x*) > *degs*(*i*) **then**
34:          *currstate* ← *INMIS*
35:          **send** *decide*(*k*) to *currneighs*
36:        **else**
37:          **send** *undecide*(*k*) to *currneighs*; *currstate* ← *UNMAK*
38:        **end if**
39:      **else if** *currstate* = *UNMAK* **then**
40:        **send** *undecide*(*k*) to *currneighs*;
41:      **end if**
42:      *phase*1*over* ← *true*
43:    **end if**
44:    **if** *phase*1*over* ∧ (*recvd*2 = *currneighs*) **then**
45:       *round_over* ← *true*; *curr_neighs* ← *curr_neighs* \ *lost_neighs*
46:       *round_recvd, phase*1*over* ← *false*; *recvd*1, *recvd*2, *lost_neighs* ← ∅
47:    **end if**
48: **end while**

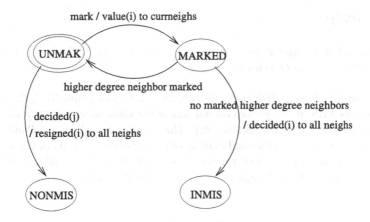

Fig. 10.6 *Rand*1_*MIS* finite-state machine

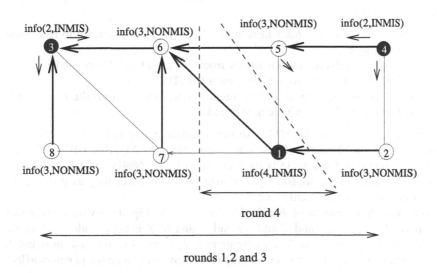

Fig. 10.7 *Rand*1_*MIS* execution example

MARKED states with higher probabilities as they have lower degree than the others. After exchanging *info* messages, however, 8 unmarks itself as it has a higher degree neighbor (3); 2 unmarks itself after tie is broken with 4 (4 > 2). In the second round, the decisions made in the previous round are communicated to neighbors with the *info*(2, *INMIS*) messages. At the end of the second round, nodes 8, 7, 6, 5, 2 resign from being in MIS, and they inform their decisions at the beginning of round 3 with the *info*(3, *NONMIS*) message. In round 4, node 1 is the only neighbor left, and it enters MIS.

10.4.1 Analysis

Theorem 10.3 *Algorithm Rand1_MIS correctly constructs a MIS of graph $G(V, E)$ in $O(\log n)$ time using $O(m \log n)$ messages.*

Proof Algorithm *Rand1_MIS* correctly constructs a MIS of graph $G(V, E)$ since if a node u joins MIS, its neighbors do not join at the same time, and they all resign from being in MIS in the consecutive step. The message complexity of *Rand1_MIS* algorithm is $O(m \log n)$ as the total number of messages exchanged at each step will be proportional to the number of edges m as shown in *Rank_MIS*, and since its time complexity is $O(\log n)$ with high probability as shown in [13]. □

10.5 The Second Random MIS Algorithm

The second randomized algorithm, called *Rand2_MIS*, also operates in synchronous rounds as follows:

1. Each node u chooses a random value $rval(v) \in [0, 1]$ and sends it to its current active neighbors.
2. Each node u collects random values from its current neighbors, and if for all $v \in \Gamma(u)$, $rval(u) > rval(v)$, node u joins the MIS.
3. If u has joined the MIS, u and its neighbors are deleted from the graph. Only active nodes continue with the next round.

The operation of *Rand2_MIS* is shown in detail in Algorithm 10.4, where we have not used an FSM, but the messages are similar to Algorithm 10.2. The termination condition for a node is its decision to be in INMIS or NONMIS state; however, it continues to transfer synchronization messages as long as it has active neighbors that have not decided yet.

An example operation of *Rand2_MIS* is shown in Fig. 10.8 with eight nodes numbered $1, \ldots, 8$. In round 1, nodes 8 and 2 enter MIS as their random values are higher than their neighbors. Their neighbors 3, 7, 1, and 4 resign, and in round 4, node 5 draws a value higher than the only other remaining node 6 and enters MIS.

10.5.1 Analysis

Theorem 10.4 *Algorithm Rand2_MIS correctly constructs a MIS of graph $G(V, E)$ in $O(\log n)$ time using $O(m \log n)$ messages.*

Proof The MIS of G produced by algorithm *Rand2_MIS* is correct using the same argument as in the *Rand2_MIS* algorithm. The message complexity of *Rand1_MIS* algorithm is $O(m \log n)$ as the total number of messages exchanged at each step is proportional to the number of edges m as shown in *Rank_MIS*, and since its time complexity is $O(\log n)$ with high probability as shown in [13]. □

Algorithm 10.4 *Rand2_MIS*

1: **set of int** *curr_neighs* ← Γ(*i*); *received, values, lost_neighs* ← ∅
2: **message types** *round, info*
3: **states** *inMIS, nonMIS, state* ← *unDEC*
4: **int** *c, r, sta*
5: **boolean** *inflag, outflag, round_recvd, round_over* ← *false*
6: **while** ¬*round_over* **do**
7:      **receive** *msg*(*j*)
8:      **case** *msg*(*j*).*type* **of**
9:                      round(*k*):   **if** *inflag* **then** *state* ← *inMIS*
10:   *inflag* ← *false*
11:                                      **else if** *outflag* **then** *state* ← *nonMIS*
12:   *outflag* ← *false*
13:                                      **if** *state* = *unDEC* **draw** *rval* ∈ [0, 1]
14:                                      **send** *info*(*k, rval, state*) to *curr_neighs*
15:                                      *round_recvd* ← *true*
16:                      info(*k, r, sta*):   *received* ← *received* ∪ {*j*}
17:                                      *values* ← *values* ∪ {*r*}
18:                                      **if** *sta* = *inMIS* **then** *outflag* ← *true*
19:                                      **if** *sta* = *inMIS*/*nonMIS*
20:   *lost_neighs* ← *lost_neighs* ∪ {*j*}
21:      **if** *round_recvd* ∧ (*received* = *curr_neighs*) **then**
22:          **if** ¬*outflag* ∧ (∀*x* ∈ *curr_neighs*: *rval* > *r_x* ∈ *values*) **then**
23:              *inflag* ← *true*
24:          **end if**
25:          *round_over* ← *true*; *curr_neighs* ← *curr_neighs* \ *lost_neighs*
26:          *round_recvd* ← *false*; *received, values, lost_neighs* ← ∅
27:      **end if**
28: **end while**

10.6 MIS Construction from Vertex Coloring

We have seen algorithms to color vertices of a graph G in Chap. 9. Each color class of the colored graph G consists of vertices colored by the same color. Since each color class is in fact an independent set as no two vertices of the same color can be adjacent, a simple way to build a MIS of G then is to start with a color class of G, include all nodes in this class in MIS, and continue to include other nodes in other classes that do not disturb the independent set property, that is, they are not neighbors with the already included nodes. Algorithm 10.5 shows a possible way to implement *Vcol_MIS*, which works for a k-colored graph G, and in each synchronous round, the member nodes of the class specified by the round number in the *round* message are included in the MIS if they are not neighbors with the nodes that are already included in the MIS.

The array *neigh_cols*[Δ] holds the number of neighbors of a node so that the xth entry has the count of neighbors with color x. Each node with color *col* at round *col* either sends a *decide* message if it can enter MIS or an *undecide* message if this

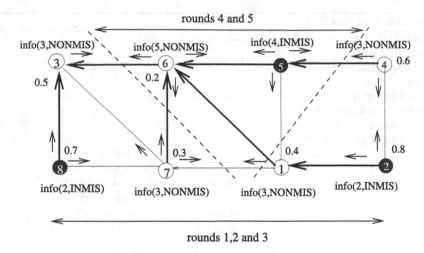

Fig. 10.8 *Rand2_MIS* execution example

is not possible due to a neighbor already in MIS. All other nodes do not send any messages until the round is their turn.

Figure 10.9(a) shows a graph with eight nodes that has a Δ of 4 and colored with four colors. In the first round of the algorithm, nodes with color 1, which are shown by double circles, include themselves in the MIS as they have no neighbors in MIS and inform their neighbors by the *decide* messages. All of their neighbors mark them as in MIS and change their state to *non_MIS*, so that they will not be involved in further message transfers. In the second round, the nodes with color 2 are adjacent to the nodes that are already in the MIS, so they do not join MIS. In the third round, however, there is a node that has color 3 and does not have a neighbor in MIS; therefore, it can enter MIS. Although the algorithm should work k rounds for a k-colored graph, MIS can be found before. In this case, additional control messages that convergecast the status of nodes to the root can be used, and when all nodes have determined their status as being in MIS or not, the algorithm terminates.

10.6.1 Analysis

Theorem 10.5 *Algorithm Vcol_MIS correctly constructs a MIS of a k-colored graph $G(V, E)$ in $O(k)$ time using $O(m)$ messages.*

Proof For a k-colored graph G, algorithm *Vcol_MIS* constructs a MIS due to the property of vertex coloring. It should be noted that all nodes of color 1 will be included in the MIS as they do not have any neighbors that are in MIS, but nodes of other color may not join MIS if they have neighbors that are already in MIS. There is a total of k rounds, and each edge of G will be traversed at most by two messages; either by a *decide* message and an *undecide* message, or two *undecide* messages, and hence the message complexity is $O(m)$.   □

Algorithm 10.5 *VCol_MIS*

```
 1: set of int received ← ∅
 2: states inMIS, nonMIS
 3: message types round, color
 4: int neigh_cols[Δ + 1], my_color
 5: boolean round_over, round_recvd, colored ← false
 6: for col = 1 to k do
 7:      { a single round for all nodes}
 8:      while ¬round_over do
 9:          receive msg(j)
10:              case msg(j).type of
11:                  round(col):    if (my_color = col) ∧ (state ≠ nonMIS)      ▷ if allowed
12:                                    send decide(col) to curr_neighs          ▷ enter MIS
13:                                    state ← inMIS
14:                                  else send undecide(col) to j
15:                                  round_recvd ← true
16:                  decide(c):     received ← received ∪ {j}
17:                                  state ← nonMIS
18:                  undecide(c):   received ← received ∪ {j}
19:
20:              if round_rcvd ∧ (received = neigh_cols[col]) then
21:                  round_over ← true
22:                  round_recvd ← false, received ← ∅
23:          end if
24:      end while
25: end for
```

10.7 Self-Stabilizing MIS Algorithms

In this section, we describe three self-stabilizing algorithms to find MIS. The first algorithm is uniform and requires a central daemon to find MIS of anonymous nodes. The second algorithm works with a distributed scheduler and finds MIS of nodes with unique identifiers. The last algorithm presented is distributed, needs node identifiers, and has the best performance.

10.7.1 Shukla's Algorithm

The first algorithm, called *Shukla_MIS*, is due to Shukla et al. [11] and considers an arbitrary graph under a central daemon with anonymous nodes. Each node has a local variable $s(i)$ that indicates after stabilization whether it is in MIS or not by applying two rules. A node joins MIS if it has no neighbors in MIS and leaves MIS if it has at least one neighbor in MIS as shown in Algorithm 10.6.

Figure 10.10 depicts an example execution of *Shukla_MIS*, where three nodes initially have marked themselves but leave MIS by Rule 2 (b), and two nodes enter MIS by Rule 1 (c) at different steps as there is a central daemon enabling execution of a single node at any step.

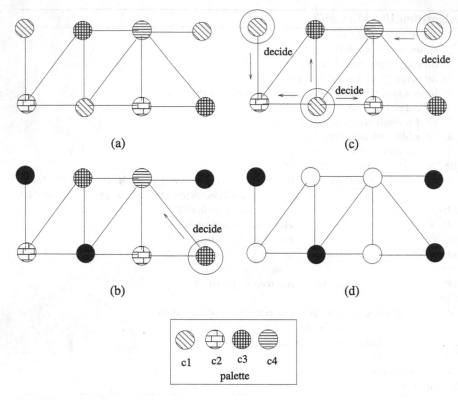

Fig. 10.9 *Vcol_MIS* execution example

Algorithm 10.6 *Shukla_MIS*

1: **Rule 1:**
2: **if** $s(i) = 0 \wedge (\forall j \in N(i))(s(j) = 0)$ **then**
3:     $s(i) \leftarrow 1$
4: **end if**
5:
6: **Rule 2:**
7: **if** $s(i) = 1 \wedge (\exists j \in N(i))(s(j) = 1)$ **then**
8:     $s(i) \leftarrow 0$
9: **end if**

10.7.1.1 Analysis

We need to show that *Shukla_MIS* stabilizes and converges in a certain number of steps and also that the output is indeed a MIS.

Lemma 10.1 *If node i executes R1 of Shukla_MIS, i and all its neighbors will be permanently disabled.*

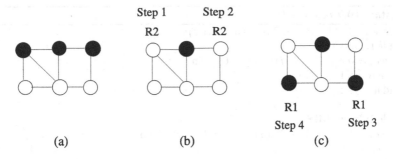

Fig. 10.10 *Shukla_MIS* execution example

Proof A node i can only execute R1 if $s(j) = 0$ for all $j \in N(i)$, and when i sets $s(i) = 1$, the neighbors cannot execute R2 as all of them have $s(j) = 0$, and they are permanently disabled. Node i cannot change its value either as all its neighbors are disabled.   □

Lemma 10.2 *Shukla_MIS stabilizes in $O(n)$ steps under a central daemon.*

Proof By Lemma 10.1, node i that has executed R1 is permanently disabled and for node i that has executed R2, we will have $s(i) = 0$, and the only rule that it can now execute is R1, and if it does so, it will be permanently disabled. Hence, any node can only have a maximum of two moves. The total number of steps required for stabilization therefore is $2n$ as there will be only one node enabled for a move at each step.   □

Theorem 10.6 *When Shukla_MIS stabilizes, the nodes will form a MIS.*

Proof We need to show that when the two rules do not apply to any node i, any node i that has $s(i) = 1$ has formed a MIS. Disabling of the two rules means that the following statement is true:

$$\neg\big((s(i) = 0) \wedge (\forall j \in N(i))(s(j) = 0)\big) \wedge \neg\big(s(i) = 1 \wedge (\exists j \in N(i))(s(j) = 1)\big)$$

$$\equiv \big(s(i) = 1 \vee \exists j \in N(i)\big)(s(j) = 0)\big) \wedge \big(\forall j \in N(i), s(i) = 0 \vee s(i) = 1\big)$$

$$\equiv \neg\big((\forall j \in N(i), s(i) = 0) \oplus (s(i) = 1)\big).$$

When all neighbors of node i are not in MIS, node i will be in MIS by R1. To check whether the IS is a MIS, we add a node i that has $s(i) = 0$ to IS, which would mean that there is $j \in N(i)$ such that $s(j) = 1$, which violates independence.   □

Shukla et al. also showed that it is impossible to obtain a deterministic and uniform algorithm for the MIS problem in an anonymous system under a distributed daemon.

Algorithm 10.7 *Ikeda_MIS*

1: $mis_i \in \{0, 1\}$ {1 if P_i is in MIS, 0 otherwise}
2: **Rule 1: Join MIS**
3: **if** $mis_i = 0 \wedge (\forall P_k \in N(i))(mis_k = 0)$ **then**
4:      $mis_i \leftarrow 1$
5: **end if**
6:
7: **Rule 2: Leave MIS**
8: **if** $mis_i = 1 \wedge (\exists P_k \in N(i))(mis_k = 1 \wedge P_k < P_i)$ **then**
9:      $mis_i \leftarrow 0$
10: **end if**

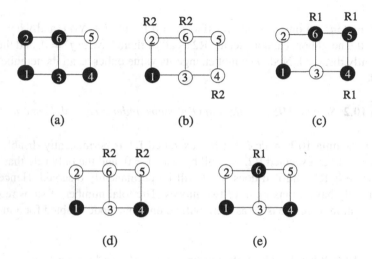

(a)   (b)   (c)

(d)   (e)

Fig. 10.11 *Ikeda_MIS* execution example

10.7.2 Ikeda's Algorithm

Ikeda et al. [4] proposed a deterministic self-stabilizing algorithm, which we will call *Ikeda_MIS*, to compute MIS under a distributed scheduler that works for arbitrary graphs and assumes that nodes have unique identifiers. Identifiers are used to break ties under a distributed daemon. Each node joins MIS if it has no neighbors in MIS and each node leaves MIS if one of its neighbors with a lower identifier is in MIS.

Formally, each process P_i has a local variable $mis_i \in \{0, 1\}$ such that $mis_i = 1$ if P_i is a member of MIS and $mis_i = 0$ if P_i is not a member of MIS. The algorithm has two rules as shown in Algorithm 10.7. R2 ensures that a node with a smaller identifier has priority to stay in MIS.

An example operation of *Ikeda_MIS* is shown in Fig. 10.11 with six nodes numbered $1, \ldots, 6$. The initial state is shown in (a) with nodes 1, 2, 3, 4, and 6 in MIS and 5 and 6 outside. Applying R2 in the first step yields nodes 2 and 3 resigning from MIS as they have a lower identifier neighbor 1 in MIS, and nodes 6 and 4 resigns similarly for neighbor 2. In the second step, nodes 4, 5, and 6 are enabled for R1, and they enter MIS. In the third step, nodes 6 and 5 resign by R2 in (c), and, finally, node 6 enters MIS by R1, and stabilization is achieved as no rules are enabled.

10.7.2.1 Analysis

Theorem 10.7 *Ikeda_MIS stabilizes in $(n + 2)(n + 1)/4$ steps.*

Proof At least one process decides its MIS value to one in each connected component, and since each node has at least one neighbor, that neighbor also decides its MIS value to 0. If graph G has c components, at least two nodes decide at each iteration, and there will be $c - 2$ nodes in the next iteration. Based on this observation, the maximum convergence time can be calculated as follows:

The number of steps

$$= (n) + (n - 2) + (n - 4) + \cdots$$
$$= 1/2\{2(n) + 2(n - 2) + 2(n - 4) + \cdots\}$$
$$< 1/2\{(n + 1) + (n) + (n - 1) + (n - 2) + (n - 3) + (n - 4) + \cdots + 1\}$$
$$= 1/2\sum_{i=1}^{n+1} i = (n + 2)(n + 1)/4.$$
$$\square$$

10.7.3 Turau's Algorithm

The third algorithm *Turau_MIS*, described in [12], uses three states as IN, OUT, and WAIT. The state IN means that the node is in MIS, and OUT indicates that the node is not in MIS. The state WAIT shows that a node wants to change into state IN; it may do so if it has no neighbor with the same state with lower identifier. In order to formally define the rules, the following predicates for a node v are needed:

- $inNeighbor(v) \equiv \exists w \in N(v) : w.state = IN.$
- $waitNeighborWithLowerId(v) \equiv \exists w \in N(v) : w.state = WAIT \wedge w.id < v.id.$
- $inNeighborWithLowerId(v) \equiv \exists w \in N(v) : w.state = IN \wedge w.id < v.id.$

These predicates are used to determine the states of nodes as shown in Algorithm 10.8. It can be seen that when node v is not part of a MIS, even when all of its neighbors are not in IN state, it attempts to enter MIS by changing its state to WAIT (*Rule 1*). In WAIT state, if one of the neighbors of v have decided to be in IN state, v changes its state to OUT (*Rule 2*). The only way for v to change its state

---

Algorithm 10.8 *Turau_MIS*

1: **Rule 1:**
2: **if** *state* = *OUT* ∧ ¬*inNeighbor*(*v*) **then**
3:     *state* ← *WAIT*
4: **end if**
5: **Rule 2:**
6: **if** *state* = *WAIT* ∧ *inNeighbor*(*v*) **then**
7:     *state* ← *OUT*
8: **end if**
9: **Rule 3:**
10: **if** *state* = *WAIT* ∧ ¬*inNeighbor*(*v*) ∧ ¬*waitNeighborWithLowerId*(*v*) **then**
11:     *state* ← *IN*
12: **end if**
13: **Rule 4:**
14: **if** *state* = *WAIT* ∧ *inNeighbor*(*v*) **then**
15:     *state* ← *OUT*
16: **end if**

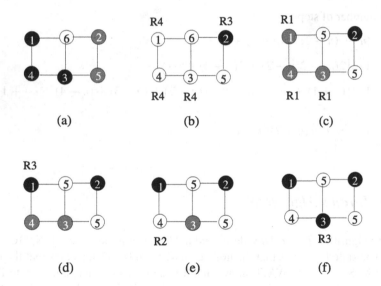

Fig. 10.12 *Turau_MIS* execution example

from WAIT to IN is that it has no neighbor *w* that is in IN state and no neighbor *w* that is in WAIT state with lower identifier (*Rule 3*). Finally, even if *v* has decided to be in IN state, it resigns by changing its state to OUT if there is a neighbor that has entered IN state (*Rule 4*).

An example operation of *Turau_MIS* is shown in Fig. 10.12 with six nodes numbered 1, ..., 6. The initial state is shown in (a) with nodes 1, 4, and 3 in MIS and 5 and 2 in WAIT state shown by gray color. Applying R4 in the first step results in nodes 1, 4, 3 changing to OUT state, and node 2 to IN state by R3 as shown in (b).

Nodes 1, 4, 3 enter WAIT state by R1 in (c), and node 1 enters IN state by R3 in (d). Node 4 goes into OUT state as it has neighbor 1, which is IN state by R2 in (e), and finally node 3 enters IN state in (f) as it has all OUT neighbors by R3. This configuration reached after 5 steps for a total of 10 moves will not change as all of the rules are disabled for all nodes.

10.7.3.1 Analysis

Lemma 10.3 *In any configuration in which all the four rules are disabled for each node, the set $I = \{v|v.State = IN\}$ is a MIS of G.*

Proof We show this by contradiction. Node v cannot have a neighbor w with state IN as w would not have entered MIS since $v.is < w.id$. Furthermore, since all the preconditions for R3 are enabled, v goes int IN state, meaning that it could not be in WAIT state. The set I is an independent set as R4 is not enabled (v has no neighbors in IN state); also it cannot be extended as R1 is not enabled. It can also be seen that once I is formed, it is attained since none of the four rules is enabled again. □

When algorithm stabilizes, an IS is obtained where each node v is either in IN or OUT state as there will be no WAIT state nodes. Since no rules will be enabled, IS is not changed again.

Theorem 10.8 *Algorithm Turau_MIS is self-stabilizing under an unfair distributed scheduler and stabilizes after at most $3n$ moves, and this bound is attained.*

Proof Assuming the case of n nodes with ascending identifiers along a line and each being in state IN, they will first change to state OUT by R4, then to state WAIT by R1, and finally to their stable state as IN if they have lower identifiers than their neighbors or OUT otherwise. Thus, each node makes three moves, and the algorithm stabilizes in $3n$ moves. □

10.8 Chapter Notes

Early work by Alon et al. [1] studied finding MIS in parallel. Kuhn et al. [5] showed that the lower bounds on time complexity for constructing a MIS are $\omega\sqrt{\log n}$ or $\omega\Delta$. MIS for special graphs has been studied by some researchers. The growth bounded graphs are investigated in [10] and trees in [6]. The best deterministic MIS algorithm has the $2^{O(\sqrt{\log n})}$ time complexity [8].

Goddard et al. [3] modified the rules of *Ikeda_MIS* such that a node joins MIS if it does not have a neighbor in MIS with higher identifier and leaves MIS if it has a neighbor with higher identifier. They also showed that the algorithm stabilizes in n rounds using the fully distributed scheduler, however, makes $O(n^2)$ moves in

Fig. 10.13 Example graph
for Exercise 2

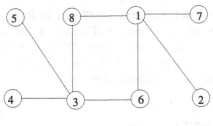

Fig. 10.14 Example graph
for Exercise 6

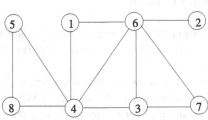

the worst case. Algorithm *Turau_MIS* is the first linear time uniform self-stabilizing
algorithm to find MIS using a distributed unfair scheduler.

Given a graph G with vertices having weights, *maximum weight independent set*
(*MWIS*) problem is finding the independent set with the largest total weight among
all independent sets of *G*, which is NP-hard. Sanghavi et al. [9] provided an algorithm to find MWIS in an arbitrary graph using *LP* relaxation. Basagni [2] provided
a linear time distributed *MWIS* algorithm for wireless networks by partitioning the
network into clusters.

10.8.1 Exercises

1. Provide necessary modifications to the algorithm *Rank_MIS* so that there are
   two phases of a round where any node that receives a *decide* message in the
   first phase sends a *resign* message in the second phase. All other nodes send
   unresign message in the second phase to provide synchronization. Provide also
   the necessary modifications to the FSM for this algorithm.
2. Using the approach of *Rank_MIS*, design a synchronous distributed algorithm
   that chooses the lowest degree node among neighbors to be included in MIS in
   each round. Show the step-by-step execution of this algorithm in the example
   graph of Fig. 10.13.
3. Provide the necessary synchronization messages to *Rand*1_*MIS* algorithm so that
   it does not always have to run *n* rounds.
4. A Depth First Search (DFS) algorithm may be used such that every time a vertex
   is visited and decided to be in MIS, the next vertex is decided not to be in MIS.
   Provide the code for a sequential DFS-based algorithm (*SeqDFS_MIS*) to find
   MIS of a graph *G*.

5. Provide a pseudocode of a distributed DFS-based MIS algorithm based on *SeqDFS_MIS* and work out its time and message complexities.
6. Provide a pseudocode distributed algorithm to find MIS of a tree *T* and work out its time and message complexities. Show the step-by-step execution of this algorithm in the example graph of Fig. 10.14 by first forming a BFS tree of the graph rooted at node 7.

References

1. Alon N, Babai L, Itai A (1986) A fast and simple randomized parallel algorithm for the maximal independent set problem. J Algorithms 7(4):567–583
2. Basagni S (2001) Finding a maximal independent set in wireless networks. Telecommun Syst 18(1–3):155–168
3. Goddard W, Hedetniemi ST, Jacobs DP, Srimani PK (2003) Self-stabilizing protocols for maximal matching and maximal independent sets for ad hoc networks. In: Proc int parallel and distributed processing symposium
4. Ikeda M, Kamei S, Kakugawa H (2002) A space-optimal self-stabilizing algorithm for the maximal independent set problem. In: Proc 3rd international conference on parallel and distributed computing, applications and technologies
5. Kuhn F, Moscibroda T, Wattenhofer R (2004) What cannot be computed locally! In: Proc 23rd ACM symposium on principles of distributed computing (PODC)
6. Lenzen C, Wattenhofer R (2011) MIS on trees. In: 30th ACM symposium on principles of distributed computing (PODC), pp 41–48
7. Luby M (1986) A simple parallel algorithm for the maximal independent set problem. SIAM J Comput 15(4):1036–1053
8. Panconesi A, Srinivasan A (1996) On the complexity of distributed network decomposition. J Algorithms 20(2):356–374
9. Sanghavi S, Shah D, Willsky AS (2009) Message passing for maximum weight independent set. IEEE Trans Inf Theory 55(11):4822–4834
10. Schneider J, Wattenhofer R (2008) A log-star distributed maximal independent set algorithm for growth-bounded graphs. In: 27th ACM symposium on principles of distributed computing (PODC)
11. Shukla SK, Rosenkrantz DJ, Ravi SS (1995) Observations on self-stabilizing graph algorithms for anonymous networks. In: Proc 2nd workshop on self-stabilizing systems
12. Turau V (2007) Linear self-stabilizing algorithms for the independent and dominating set problems using an unfair distributed scheduler. Inf Process Lett 103(3):88–93
13. Wattenhofer R Principles of distributed computing. Lecture notes, ETH

Chapter 11
Dominating Sets

Abstract A subset of the vertices of a graph is a dominating set if every vertex not in the subset is adjacent to at least one vertex in this subset. Dominating sets are widely used for clustering and routing in ad hoc wireless networks. In this chapter, we describe sample sequential, distributed, and self-stabilizing dominating set algorithms.

11.1 Introduction

A dominating set can be defined formally as follows.

Definition 11.1 Given a graph $G(V, E)$, a *Dominating Set* (*DS*) is the set of vertices $V' \in V$ such that any vertex $v \in V$ is either in V' or a neighbor of a vertex in V'.

Alternatively, for all $v \in (V - V')$, v is a neighbor to at least one node in V'. Therefore, every MIS is a DS. However, every DS is not a MIS since some vertices of a DS may be neighbors.

Definition 11.2 (Minimum and minimal dominating sets) A dominating set is minimum (*MinDS*) if it has the smallest cardinality among all possible dominating sets of the graph G. A dominating set is minimal (*MDS*) if it is not contained in any other dominating sets of G.

Finding a minimum sized dominating set is NP-hard [2]. If all nodes in a dominating set are connected, that is, there is a path between any pair of vertices in the set, it is called a *connected dominating set*. A formal definition is given below.

Definition 11.3 (Connected Dominating Set) A *Connected Dominating Set* (*CDS*) of a graph G is a dominating set that induces a connected subgraph in G.

A simple method to construct a CDS would involve finding the MIS of the graph and then include additional vertices to connect the vertices in the MIS.

The *Minimum Connected Dominating Set* (*MinCDS*) is a connected dominating set with the minimum size, and finding MinCDS is NP-hard. The *Minimal Connected Dominating Set* (*MCDS*) is a CDS that is not contained in any other CDS

K. Erciyes, *Distributed Graph Algorithms for Computer Networks*,
Computer Communications and Networks, DOI 10.1007/978-1-4471-5173-9_11,
© Springer-Verlag London 2013

Fig. 11.1 Dominating set
examples

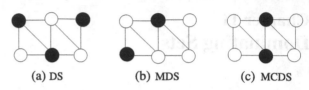

(a) DS   (b) MDS   (c) MCDS

of G. The *Weakly Connected Dominating Set (WCDS)* is a dominating set where
DS and the neighbors of nodes in the DS induce a connected subgraph of G. Com-
puting WCDS is NP-hard, and in a connected graph, all dominating sets are WCDS.
Figure 11.1 displays dominating set examples where (a) is a DS of size 3, (b) is a
MinDS of size 2, and (c) is a MinCDS of size 2.

As finding MinCDS is NP-hard, approximation algorithms that achieve convergence to optimum (*OPT*) value for MinCDS are used. In this chapter, we first describe sequential algorithms to find MCDS and then show distributed algorithms
based on these sequential algorithms to form MCDS. We conclude by sample self-
stabilizing algorithms that produce DS and CDS.

11.2 Sequential Algorithms

We will describe four sequential algorithms to find minimal dominating sets in this
section. All these algorithms use simple heuristics to find MDS or MCDS.

11.2.1 Greedy Sequential MDS Algorithm

Since our aim is to have a dominating set as small as possible, we should greedily
search for nodes that dominate as many neighbors as possible. The first greedy sequential algorithm finds an MDS by always selecting nodes that have the greatest
number of neighbors that do not dominate or are not dominated. We will assume
that the nodes in the MCDS are colored black, their neighbors are colored gray, and
any node that has not been a dominator or dominated is colored white.

The *span* of a node is defined as the number of white nodes it has, including
itself. The *Seq_MDS* algorithm initially colors all vertices of the graph white. At
each iteration it then picks the vertex u with the highest span and colors it black to
include it in the DS. It also colors all the neighbors of u gray in this iteration, and
vertex u is then excluded from the search list. From the remaining vertices the one
with the highest span is selected in the next iteration. This process is continued until
there are no white colored vertices left. Algorithm 11.1 shows a possible detailed
implementation of this algorithm, where the array *spans* holds the current spans of
all nodes, set S holds the white nodes, and the array *colors* has the current colors of
all nodes, and the MDS set is the output of the algorithm.

Figure 11.2 shows the execution of *Seq_MDS* algorithm in example graph with
10 nodes. Node 4 has the highest span and is included in MDS, and all its neighbors

Algorithm 11.1 *Seq_MDS*

1: Input: $G(V, E)$
2: S: set of white nodes at any time
3: *spans*[n]: array holding spans for every node
4: *colors*[n]: array holding colors of nodes
5: *MDS*: returned dominating set
6: $MDS \leftarrow \varnothing$
7: **for all** $v \in V$ **do**   ▷ initialize all vertices to white
8:     *colors*[v] $\leftarrow$ *white*
9:     *spans*[v] $\leftarrow |\Gamma(v)| + 1$ ▷ initialize spans
10: **end for**
11: $S \leftarrow V$
12: **while** $S \neq \varnothing$ **do**   ▷ Loop until no more white vertices left
13:     $u \leftarrow \max\{w | spans[w]\}$ ▷ find vertex u with max span
14:     **for all** $v \in \Gamma(u)$ **do**
15:         **if** *colors*[v] $=$ *white* **then**
16:             **for all** $w \in \Gamma(v)$ **do**   ▷ decrement spans of white neighbors
17:                 *spans*[w] $\leftarrow$ *spans*[w] $- 1$
18:             **end for**
19:             *colors*[v] $=$ *gray*
20:         **end if**
21:         **if** *colors*[u] $=$ *white* **then**
22:             *spans*[v] $\leftarrow$ *spans*[v] $- 1$
23:         **end if**
24:     **end for**
25:     *colors*[u] $\leftarrow$ *black*
26:     *spans*[u] $\leftarrow 0$
27:     $MDS \leftarrow MDS \cup \{u\}$ ▷ insert the highest span vertex in MDS
28:     $S \leftarrow S \setminus \{\Gamma(u) \cup \{u\}\}$ ▷ remove vertex and neighbors from S
29: **end while**

are colored gray as shown in (a). The node with the highest span in the next iteration is 3, which is colored black, and its neighbors are colored gray as shown in (b). The third iteration is displayed in (c), where node 8, which has the same span as node 5, is selected based on the magnitude of identifiers. The final MDS is shown in (d), where node 5 is colored black, and finally all nodes are either colored black or gray.

The algorithm correctly finds an MDS as all the nodes will end up as black or gray in the end. The approximation ratio of this algorithm is $\ln \Delta$ as shown in [10], and the total number of iterations will be $O(n)$.

11.2.2 Greedy Sequential MCDS Algorithm

The second version of the greedy algorithm finds an MCDS by a simple modification. Instead of choosing any vertex with the highest span, whether it is a white or

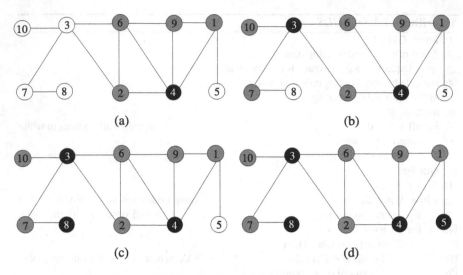

Fig. 11.2 The sequential greedy MDS algorithm example

gray node, we choose a gray vertex with the highest span. As a gray node will be adjacent to a black node, this algorithm will add a black node next to at least one other black node in each iteration; therefore, the output will be an MCDS. We will only show the execution of this algorithm in Fig. 11.3, where a network of 10 nodes is numbered 1, . . . , 10. Nodes 4 and 1 have the highest spans, and the tie is broken by the identifiers so that node 4 is elected as the first node to be colored black and all of its neighbors are colored gray. The algorithm than selects node 3 to color as it is gray and has the highest number of white neighbors. This node is colored black, and all its neighbors gray as shown in (b). In the third iteration, node 1 is selected as it is the only gray node with white neighbors to give the MCDS consisting of nodes 4, 3, and 1 as shown in (c). This algorithm also has the time complexity $O(n)$.

11.2.3 Guha–Khuller Algorithms

Guha and Khuller [4] provided two greedy sequential algorithms to construct CDS in arbitrary graphs. In their first algorithm, a CDS is grown iteratively from a single node. In the second algorithm, a WCDS is first formed, and then the intermediate nodes are connected. These algorithms have formed the basis for other distributed MCDS algorithms, and we will describe them next.

11.2.3.1 The First Algorithm

This algorithm called *GK1_CDS* provides an improvement over the *Seq_CDS* by the addition of a simple heuristic. It starts by coloring all the vertices of the graph

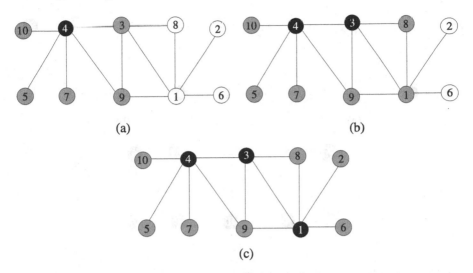

Fig. 11.3 The sequential greedy MCDS algorithm example

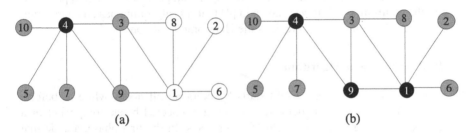

Fig. 11.4 Guha–Khuller first algorithm example

$G(V, E)$ white, and it first selects the vertex with the highest degree, colors it black, and all its neighbors are colored gray. The algorithm then scans all the gray nodes and their white neighbors by finding the number of white neighbors of gray nodes and the white neighbors of their neighbors. It then selects a gray node or a pair consisting of a gray node and a white node, whichever gives the highest number of white neighbors. It colors the selected gray node or the selected pair of gray and white node black and starts with the next iteration. This process continues until all the nodes are colored black or gray. As before, the black nodes will form the CDS.

Figure 11.4 displays the operation of this algorithm in the same sample graph of Fig. 11.3. Node 4 is first selected; it is colored black, and all of its neighbors are colored gray as shown in (a). Then, all gray neighbors and their white neighbors are searched, and the node pair ⟨9, 1⟩ (or the pair ⟨3, 1⟩) has the highest number of white neighbors, both are colored black, and their neighbors are colored gray as shown in (b), which completes the construction of the CDS consisting of nodes 4, 9, and 1.

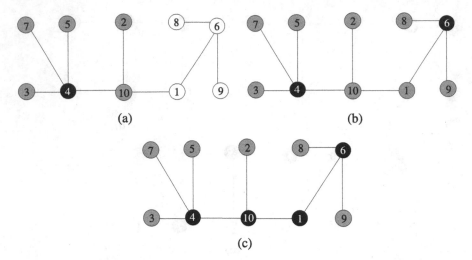

Fig. 11.5 Guha–Khuller second algorithm example

The simple modification provided by this algorithm provides a better approximation to the optimum CDS. It was shown in [4] that this algorithm has an approximation ratio of $2(1 + H(\Delta))$, where H is the Harmonic Function.

11.2.3.2 The Second Algorithm

The second algorithm (*GK2_CDS*) starts by coloring all nodes white as before.
A *piece* in this algorithm is defined as either a connected black component or a
white node. The algorithms consists of two phases. In the first phase, a node that
causes the greatest reduction in the number of pieces in the graph is selected and
colored black, and all of its neighbors are colored gray. At the end of the first phase,
each node in the graph is either colored gray or black. The second phase uses a
Steiner Tree algorithm to connect the black nodes so that an MCDS is formed [4].

This algorithm yields a CDS with an approximation ratio of $3 + \ln \Delta$ [4]. Figure 11.5 displays the operation of this algorithm in a graph with 10 nodes. Node 4 is
colored black as it has the highest degree, and all of its neighbors are colored gray
initially. Node 6 is colored black, and its neighbors gray next, as it provides the maximum reduction in the number of pieces as shown in (b), where all the nodes of the
graph are colored either gray or black, and a dominating set is obtained. However,
this set is not connected. In (c), nodes 10 and 1 are used to connect the dominating
set nodes of 4 and 6, resulting in the MCDS consisting of nodes 4, 10, 1, and 6.

11.3 Distributed Algorithms

In this section, a greedy algorithm to construct an MDS and another similar algorithm to construct an MCDS are described.

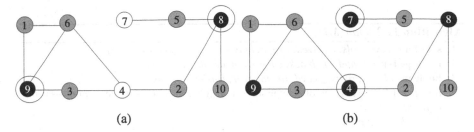

Fig. 11.6 Greedy MDS algorithm example

11.3.1 Greedy MDS Algorithm

We will first explore the possibility of obtaining a distributed algorithm from the sequential algorithm *Seq_MDS*, where a node with the highest span is colored black in each iteration. In the synchronous implementation, a node that finds it has the highest span among neighbors, colors itself black, and the neighbors color themselves gray in each round. This process continues until there are no white nodes left.

The messages used are described below:

- *ch_black*: A node that has the highest span locally, sends this message to current active (white or gray) neighbors.
- *undecide*: A node that does not have the highest span locally, sends this message to current neighbors.
- *ch_gray*: A node that changes its color from white to gray sends this message to current active neighbors so that they decrement their spans.
- *no_change*: If a white node does not change its color in a round, it sends this message to current active neighbors.

The *undecide* and *no_change* messages are needed for synchronization. In the design we propose, each round consists of two phases. In the first phase, all active nodes exchange the *ch_black* and *undecide* messages. Any node i that changes its color to gray due to the reception of *ch_black* should inform its current neighbors, and it does so in the second phase of the round. A node that has white neighbors waits to receive either *ch_gray* or *no_change* messages from them so that the second phase can be completed. Algorithm 11.2 (*Span_MDS*) shows a way to implement the algorithm described.

Figure 11.6 shows an example network that executes the *Span_MDS*. There are 10 nodes, the nodes 9 and 8 have the highest spans locally, and they color themselves black. Nodes 9, 6, and 4 have the same spans, but node 9 has the highest identifier, so it can decide to be in MDS. Node 4 cannot color itself black because node 6 has a higher identity, which is also excluded by node 9. All the neighbors of these nodes are colored gray in the first round as shown in (a), and the spans of their neighbors 4 and 7 are decremented to be unity. In the second round, nodes 4 and 7 have the highest spans locally, and they are colored black to result in the MDS shown in (b).

Algorithm 11.2 *Span_MDS*

1: **set of int** *recvd_cols, received, lost_neighs* ← ∅; *curr_neighs* ← $\Gamma(i)$
2: **message types** *round, ch_black, ch_gray, undecide, no_change*
3: **boolean** *colored, round_recvd, round_over, finished* ← *false*
4: **int** *spans*[d_i] ← degrees of neighbors + 1, *neigh_cols*[d_i] ← whites, *color* ← *white*,
   n_recvd
5: {round *k* for all nodes}
6: **while** ¬*round_over* **do**
7:     **receive** *msg*(j)
8:         **case** *msg*(j).*type* **of**
9:             <u>*round*(k):</u>   **if** *spans*[i] ≠ 0 **then**
10:                  .   **if** *spans*[i] > *max*{j|*spans*[j]} **then**
11:   *color* ← *black*, *spans*[i] ← 0
12:   **send** *ch_black* **to** *curr_neighs*
13:                                    **else send** *undecide* **to** *curr_neighs*
14:                                *round_recvd* ← *true*
15:             <u>*ch_black*(k):</u>   *received* ← *received* ∪ {j}
16:                                *recvd_cols* ← *recvd_cols* ∪ {*black*}
17:                                **if** *neigh_cols*[j] = *white*
18:                                    *spans*[i] ← *spans*[i] − 1
19:                                *lost_neighs* ← *lost_neighs* ∪ {j}
20:                                *neigh_cols*[j] ← *black*
21:             <u>*undecide*(k):</u>   *received* ← *received* ∪ {j}
22:             <u>*ch_gray*:</u>   *n_recvd* ← *n_recvd* + 1
23:                                *spans*[i] ← *spans*[i] − 1
24:                                *neigh_cols*[j] ← *gray*
25:             <u>*no_change*:</u>   *n_recvd* ← *n_recvd* + 1
26:
27:     **if** (¬*finished* ∧ *round_recvd*) ∧ (*received* = *curr_neighs*) **then**
28:         **if** ∃*col* ∈ *recvd_cols* : *col* = black **then**
29:             **if** *color* = *white* **then**
30:                 **send** *ch_gray* **to** *curr_neighs*
31:                 *color* ← *gray*
32:             **else send** *no_change* **to** *curr_neighs*
33:             **end if**
34:         **else send** *no_change* **to** *curr_neighs*
35:         **end if**
36:         *finished* ← *true*
37:     **end if**
38:     **if** (*finished*) **then**
39:         **if** *n_recvd* = |*curr_neighs*| **then**
40:             *round_over* ← *true*; *curr_neighs* ← *curr_neighs* \ *lost_neighs*; *n_recvd* ← 0
41:             *round_recvd, finished* ← *false*; *received, recvd_cols, lost_neighs* ← ∅
42:         **end if**
43:     **end if**
44: **end while**

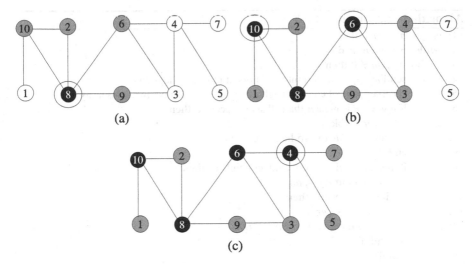

Fig. 11.7 Greedy MCDS algorithm example

11.3.1.1 Analysis

Theorem 11.1 *The Span_MCDS algorithm correctly constructs an MCDS in $O(n)$ rounds using $O(nm)$ messages.*

Proof The algorithm continues until all nodes are colored either black or gray, also removal of a black node would result in at least one neighbor to be colored white. Therefore, the output is an MDS. In the case of a linear network, there would be $\lceil n/2 \rceil$ rounds, therefore, the total number of rounds would be $O(n)$. There will be a constant time of edge traversals by the messages in each round resulting in $O(nm)$ messages altogether. $\square$

11.3.2 Greedy MCDS Algorithm

We will now implement a greedy distributed algorithm based on *Seq_MCDS* that finds MCDS of a network. The pseudocode of this algorithm will be similar to *Span_MDS* algorithm with the difference that only a gray node with the highest span locally may be selected to be colored black. Since we always choose a gray node that is adjacent to a black node to color black, the MDS is connected. Figure 11.7 shows an example network that executes the *Span_MCDS*. There are 10 nodes, and node 8 has the highest span based on identifiers which colors itself black and is included in the MCDS. All of its neighbors are colored gray in the first round. Nodes 10 and 6 are the gray nodes having the highest spans among their active neighbors and are included in the MCDS in the second round. Finally, node 4 is included to give the MCDS as shown in (c). This algorithm should start by the node that has

---

Algorithm 11.3 *Twospan_CDS*

1: *my_span* ← |Γ(*i*)| + 1
2: **while** ¬*roundover* **do**
3:      **if** *my_span* ≠ 0 **then**
4:          **send** *my_span* to current neighbors at most two hops away
5:          **receive** spans of current neighbors which are at most two hops away
6:          **if** *my_span* is greater than all spans received **then**
7:              *color* ← black
8:              **send** *decide* to two-hop neighbors
9:          **end if**
10:         **if** *decide* received from a direct neighbor **then**
11:             **decrement** *my_span*
12:             **if** *color* = white **then**
13:                 *color* ← gray
14:                 **send** *ch_gray* to *curr_neighs*
15:             **end if**
16:         **end if**
17:         **if** *ch_gray* received from a direct neighbor **then**
18:             **decrement** *my_span*
19:         **end if**
20:     **end if**
21: **end while**

---

the highest span that colors itself black; therefore, the node with the highest span should be known globally. The time and message complexities of this algorithm are $O(n)$ and $O(nm)$ as in the *Span_MDS* algorithm.

11.3.3 The Two-Span MDS Algorithm

Based on *GK1_CDS*, a distributed algorithm can be designed. Each node in this algorithm compares its span with the spans of the two-hop neighbors, that is, neighbors that are at most two hops away [10]. If a node finds that it has the highest span among its two-hop neighbors, it decides to enter the MDS. It has a similar structure to *Span_MDS* algorithm, so we will just briefly show the pseudocode as Algorithm 11.3 (*Twospan_CDS*). The statements between lines 10 and 19 require a second phase within a round as in *Span_MDS* algorithm, or odd- and even-numbered rounds can be used.

Theorem 11.2 *Algorithm* 11.3 *provides a CDS with the approximation ratio* ln Δ *to MCDS in* $O(n)$ *rounds.*

Proof The approximation ratio is similar to the *Seq_MCDS*. Since at least one node is added to the *CDS* in each round, the algorithm executes $O(n)$ rounds.   □

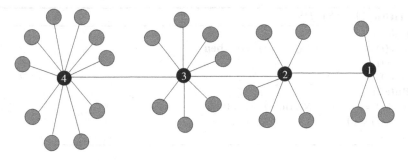

Fig. 11.8 Sequential execution of two-span MDS algorithm

Figure 11.8 displays the operation of this algorithm in a sample graph with nodes of decreasing degrees. Each node has to wait for its left neighbor node to be included in the CDS. Node 1 cannot join CDS as its left neighbor 3 has a larger span. Similarly, nodes 3 and 2 delay joining the CDS, and only node 4 can join CDS in the first round after span messages are exchanged. This is followed by nodes 3, 2, and 1 in the consecutive three rounds, resulting in the optimal CDS shown by black nodes. The execution in this case is sequential.

11.4 Self-Stabilizing Domination

In this section, we will describe self-stabilizing algorithms to find a dominating set and a minimal dominating set.

11.4.1 Dominating Set Algorithm

The first algorithm called *SS1_DS* is due to Hedetiemi et al. [6] and works under a central scheduler to find two dominating partitions of a graph. In this algorithm, each node i has a binary variable $x(i)$. When the network stabilizes, each node i having the value $x(i) = 0$ or $x(i) = 1$ belongs to dominating sets. The algorithm works with two simple rules. In Rule 1, if each neighbor j of a node i has $x(j) = 0$, then node i assigns its variable $x(i)$ to 1. Rule 2 works similarly, and this time if each neighbor j of a node i has $x(j) = 1$, then this node assigns its variable $x(i)$ to 0 as shown in Algorithm 11.4.

Figure 11.9 shows the operation of this algorithm in a simple graph of six nodes, where two nodes make moves with R1 and R2 consecutively to obtain two separate dominating sets shown by black and white colors.

Theorem 11.3 *The algorithm stabilizes in $O(n - 1)$ moves in a connected network.*

Algorithm 11.4 *SS1_DS*

1: **Rule 1:**
2: **if** $x(i) = 0 \wedge (\forall j \in N(i))(x(j) = 0)$ **then**
3:     $x(i) \leftarrow 1$
4: **end if**
5: **Rule 2:**
6: **if** $x(i) = 1 \wedge (\forall j \in N(i))(x(j) = 1)$ **then**
7:     $x(i) \leftarrow 0$
8: **end if**

Fig. 11.9 *Hedet_DS* algorithm example

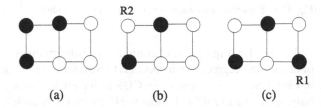

$$(a) \qquad\qquad (b) \qquad\qquad (c)$$

Proof In both rules, node i can only move when all its neighbors have the same value as itself. Therefore, whenever i moves, its value will be different from the values of its neighbors after which neither i nor its neighbors can make a move. When a node moves, the remaining nodes can move at most once, so that the total number of moves cannot exceed $n - 1$ [6].   □

11.4.2 Minimal Dominating Set Algorithm

The self-stabilizing algorithm which works under a central scheduler to find the MDS of a graph is also due to Hedetiemi et al. [6]. Every node i in this algorithm called *SS2_MDS* has a binary variable $x(i)$, which is equal to 1 if the node is in the minimal dominating set S. Each node also has a pointer ($\rightarrow$), which points to its dominator. If this pointer has null value, either $i \in S$ or i is not dominated. This algorithm is based on the fact that a dominating set DS is a minimal dominating set if and only if it is dominating and every $u \in$ DS has a private neighbor [6]. Therefore, this algorithm attempts to form a DS with each element of DS dominating only one element.

The membership rules M1 and M2 of this algorithm are described as follows:

- M1: If node i and all its neighbors are not in MDS, i enters MDS.
- M2: If node i is in MDS but has at least one neighbor that is also in CDS; and also it is not a dominator of any of its neighbors, it leaves MDS.

The pointer moves P1, P2, and P3 do not change the membership of a node in the MDS, but they modify the value of the pointers as follows:

- P1: If a node is in MDS and its pointer is not *null*, its pointer is set to *null*.

Algorithm 11.5 *SS2_DS*

1: **M1:**
2: **if** $x(i) = 0 \wedge (\forall j \in N(i))(x(j) = 0)$ **then**
3:     $x(i) \leftarrow 1$
4: **end if**
5: **M2:**
6: **if** $x(i) = 1 \wedge (\nexists j \in N(i)) \wedge (j \rightarrow i) \wedge (\exists k \in N(i))(x(k) = 1)$ **then**
7:     $x(i) \leftarrow 0$
8: **end if**
9: **P1:**
10: **if** $x(i) = 1 \wedge (i \nrightarrow null)$ **then**
11:     $i \leftarrow null$
12: **end if**
13: **P2:**
14: **if** $x(i) = 0 \wedge (\exists \text{ unique } j \in N(i))(x(j) = 1) \wedge (i \nrightarrow j))$ **then**
15:     $i \rightarrow j$
16: **end if**
17: **P3:**
18: **if** $x(i) = 0 \wedge (\exists \text{ more than one } j \in N(i))(x(j) = 1) \wedge (i \nrightarrow \varnothing))$ **then**
19:     $i \rightarrow \varnothing$
20: **end if**

- P2: If node i is not in MDS and has exactly one neighbor j in MDS, which i is not pointing, it sets its pointer to j.
- P3: If node i that is not in MDS has more than one neighbor in MDS and its pointer is not *null*, it sets its pointer to null.

Algorithm 11.5 shows a pseudocode of this algorithm.

Theorem 11.4 *The algorithm finds an MDS in $O(n^2)$ moves.*

Proof If a node moves by M1, it will not make any other membership move. If it moves by M2, the next membership move will have to be M1, and there will not be any membership moves after this move. Therefore, a node can make at most two membership moves. Also, there can be at most n consecutive pointer moves as any pointer move by a node i results in i being unprivileged, and i cannot be privileged by any pointer moves of the neighbors. Hence, each node can make one pointer move resulting in n consecutive pointer moves at most. There can be at most $2n$ membership moves. As there can be at most n pointer moves before or after each membership move, total number of moves before stabilization is $O(n^2)$ moves. A more detailed description of the proof can be found in [6]. $\square$

11.5 Chapter Notes

Finding minimal (connected) dominating sets is a well-studied topic in both graph theory and ad hoc wireless networks. There is in fact a book devoted to this prob-

Fig. 11.10 Example graphs
for Exercise 1

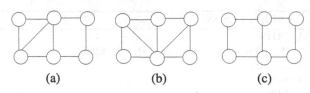

                (a)   (b)   (c)

Fig. 11.11 Example graph
for Exercise 2

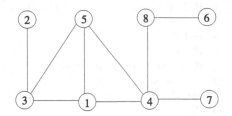

lem [5]. Dominating set problem is equivalent to *set-covering problem*. Given a set
of elements called the universe and sets whose union comprises the universe, the set
cover problem is to find the smallest number of sets whose union contains all elements in the universe. The decision version of this problem is: given (U, S), where
U is the universe, and S is a set of subsets, for an integer k, find if a set covering of
size k or less exists. This problem is shown to be NP-complete in [9].

Weighted dominating set problem is to find a minimal weight dominating set in a
graph where nodes have weights, which is also NP-complete. Chvatal [1] proposed a
central weighted-set cover-based dominating-set algorithm (CENTSET) with ln W
approximation ratio, where W is the minimum weight of the dominating set. The
dominator with the minimum weight ratio is chosen, it is covered with its neighbors
in each round, and the algorithm continues until all nodes are either dominators or
are dominated. In the self-stabilizing approach, algorithms to find CDS were given
in [7, 8]. Goddard et al. [3] recently provided a self-stabilizing algorithm to find a
connected dominating set in an anonymous network.

An MCDS can be conveniently used for routing in ad hoc networks, and we will
be investigating the construction of a backbone for routing using MCDS in ad hoc
wireless networks in Chap. 15.

11.5.1 Exercises

1. Find examples of DS in (a), MDS in (b), and MCDS in (c) of the graph of
   Fig. 11.10.
2. Provide a pseudocode for the *Seq_MCDS* algorithm and show its execution in
   the graph of Fig. 11.11.
3. Sketch a pseudocode of a greedy distributed algorithm that finds a maximal
   weighted dominating set. Work out its time and message complexities. Also,
   show the operation of this algorithm step-by-step in the example graph of
   Fig. 11.12 where each vertex is labeled by its weight.

Fig. 11.12 Example graph
for Exercise 3

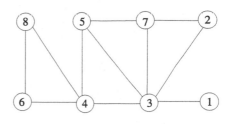

Fig. 11.13 Example graph
for Exercise 4

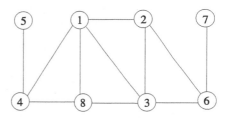

4. Provide the FSM-based version of *Span_MCDS* algorithm described in
   Sect. 11.3.1 that finds the MCDS by greedily selecting the node with the highest
   span. Provide its FSM diagram and a pseudocode and work out the time and
   message complexities. Show also step-by-step execution of this algorithm in the
   example graph of Fig. 11.13.
5. Provide a pseudocode for the SYNSET algorithm described in Sect. 11.5.

References

1. Chvatal V (1979) A greedy heuristic for the set-covering problem. Math Oper Res 4(3):233–
   235
2. Garey MR, Johnson DS (1978) Computers and intractability: a guide to the theory of NP-
   completeness. Freeman, San Francisco
3. Goddard W, Srimani P (2010) Anonymous self-stabilizing distributed algorithms for connected dominating set in a network graph. In: Proc the international multi-conference on complexity, informatics and cybernetics (IMCIC)
4. Guha S, Khuller S (1998) Approximation algorithms for connected dominating sets. Algorithmica 20(4):374–387
5. Haynes TW, Hedetniemi ST, Slater PJ (1998) Fundamentals of domination in graphs. Dekker,
   New York
6. Hedetniemi SM, Hedetniemi ST, Jacobs DP, Srimani PK (2003) Self-stabilizing algorithms
   for minimal dominating sets and maximal independent sets. Comput Math Appl 46(5–6):805–
   811
7. Jain A, Gupta A (2005) A distributed self-stabilizing algorithm for finding a connected dominating set in a graph. In: Proc PDCAT. IEEE Computer Society, Los Alamitos, pp 615–619
8. Kamei S, Kakugawa H (2007) A self-stabilizing distributed approximation algorithm for the
   minimum connected dominating set. In: IPDPS. IEEE Press, New York, pp 1–8
9. Karp RM (1991) Probabilistic recurrence relations. In: Proc 23rd annual ACM symposium on
   theory of computing (STOC 91), pp 190–197
10. Wattenhofer R Principles of distributed computing, Chapter 12. Class notes, ETH Zurich

Chapter 12
Matching

Abstract A matching M of a graph $G(V, E)$ is a subset of its edges such that no two edges in M have common endpoints. Matching is a fundamental problem in graph theory, and although there are many sequential algorithms for matching, the distributed algorithms have begun to receive attention recently due to many applications of matchings in distributed systems such as mobile and sensor networks. Matching algorithms in distributed systems may also be the building blocks for other algorithms or protocols. In this chapter, we describe sample distributed algorithms for matching in graphs.

12.1 Introduction

The formal definitions for matching-related concepts are as follows.

Definition 12.1 (Matching) Given an undirected graph $G(V, E)$, a matching (M) of G is a subset of the edges E such that no vertex in V is incident to more than one edge in M.

Definition 12.2 (Maximal matching) A *maximal matching* (MM) of a graph $G(V, E)$ is a matching of G that is not properly contained in any other matching. In other words, a matching is maximal if we cannot add any edge to the existing set.

Definition 12.3 (Maximum matching) A *maximum matching* $(MaxM)$ for a graph $G(V, E)$ is a matching of G that has the largest number of edges among all possible matchings of G.

The maximum matching problem is to find a matching in G with maximum cardinality.

Definition 12.4 (Maximum weighted matching) A *maximum weighted matching* $(MaxWM)$ for a graph $G(V, E, w)$, where $E : w \rightarrow \mathbb{R}$, is a weighted matching of G such that there does not exist any other weighted matching of G with total weight larger than MaxWM.

K. Erciyes, *Distributed Graph Algorithms for Computer Networks*,
Computer Communications and Networks, DOI 10.1007/978-1-4471-5173-9_12,
© Springer-Verlag London 2013

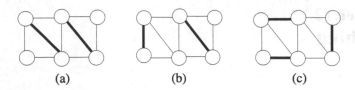

(a)   (b)   (c)

Fig. 12.1 Matching examples

Figure 12.1 displays matching examples where (a) and (b) are maximal matchings of size 2 and (c) is a MaxM of size 3.

There may be diverse applications of matching in computer networks. For example, there could be k available resources, and m events may be triggered at any time in a sensor network, and the requirement would be the handling of an event by just one node, which requires matching of m events to k sensors. A network switch that connects its input ports to output ports so that no port is connected to more than one port is an example of *bipartite matching*, where the vertex set V of graph $G(V, E)$ is partitioned into two disjoint sets V_1 and V_2, any $e \in E$ has two endpoints, one in V_1 and the other in V_2, and no $v \in V$ is incident to more than one edge. *Maximal bipartite matching* seeks to find a maximal matching in a bipartite graph. *Maximal bipartite weighted matching* for the network switch example may be required if weights are associated with packets for their sizes or priorities, and the aim would be to match the input ports of the switch to the output ports to give the maximal matching for best performance.

A *perfect matching* (1-*factor matching*) matches all vertices of the graph, in which case every vertex of the graph is incident to exactly one edge of the matching. For a matching M, an *alternating path* is a path in which the edges alternatively belong and do not belong to the matching. *An augmenting path* is an alternating path that starts from and ends on unmatched vertices. It can be proven that a matching is maximum if and only if it does not have any augmenting paths [3].

Unlike other graph problems, we have seen such as vertex coloring, maximal independent set, and minimal dominating set; maximal matching and weighted matching of a graph can be found in polynomial time using an algorithm of Edmonds [3]. In this chapter, we classify and analyze distributed matching algorithms in three sections as unweighted, weighted, and self-stabilizing algorithms.

12.2 Unweighted Matching

Unweighted matching algorithms assume that the edge weights are identical and therefore are not considered as an input parameter to the algorithm. In this section, we first describe a sequential algorithm to find a matching and then show two distributed algorithms that work in synchronous rounds to find maximal matching.

Algorithm 12.1 *Seq_MM*

1: Input $G(V, E)$
2: $S \leftarrow E, MM \leftarrow \varnothing$
3: **while** $S \neq \varnothing$ **do**
4:     **select** any $\{u, v\} \in S$
5:     $MM \leftarrow MM \cup \{u, v\}$
6:     **delete** all edges incident to either u or v from S
7: **end while**

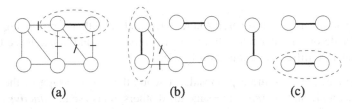

(a)   (b)   (c)

Fig. 12.2   The sequential matching algorithm example

12.2.1 A Sequential Algorithm

A sequential algorithm to find the MM arbitrarily chooses an edge e from the un-marked edge set S, includes e in the matching, and removes all edges incident to the vertices on the end points of e from the unmarked set. The algorithm *Seq_MM* continues until there are no edges left in the set S as shown in Algorithm 12.1.

In Fig. 12.2, the operation of *Seq_MM* is shown in a sample graph with the removed edges shown as marked at each iteration. As there will be at least a single edge removal at each step, the time complexity of *Seq_MM* is $O(m)$ steps.

12.2.2 The Greedy Distributed Algorithm

In our first attempt to have a distributed matching algorithm, we will again use the identifiers of nodes to make matching decisions. The nature of matching requires three distinct steps before a matching decision can be made. First, a node may re-ceive more than one offer to be matched. It should accept one of these offers and reject the others in the second step. In the third step, a node that learns the result of its proposal should inform its neighbors so that it may or may not be included in future matching decisions. This three-step decision making requires careful consid-eration of the structure of the algorithm. In the solution we propose, we have odd- and even-numbered rounds, where proposals are made in odd rounds, while accep-tance or rejection messages are sent and the informing of the neighbors is done in even rounds. Since there are two steps in an even round, synchronization is provided by using flag variables. The messages used in this algorithm are *propose, unpropose, ack*, *reject*, *neigh_ack*, and *neigh_rej*. The following describes the actions in an odd-numbered round of the algorithm.

1. If node i finds that it has the highest identifier among its active neighbors, it randomly selects an active neighbor j and proposes to this neighbor by the *propose* message.
2. All nodes other than the highest identifier node send the *unpropose* messages to their active neighbors. The sum of all messages received as *propose* and *unpropose* should be equal to the number of active neighbors for this round to finish.

At the end of an odd-numbered round, each node is either a proposing node, a proposed node or an other node. In an even-numbered round, these three types of active node have different actions as follows:

1. If node i has two or more proposals, it sends the *ack* message to the highest identifier node and the *reject* message to all others. It also sends *inactive* message to the rest of current neighbors.
2. Node i that receives an acknowledgement to its proposal made in the previous round is matched. It should also inform its neighbors that it is matched so that they do not send any proposals to it in future rounds. It achieves this by sending the *neigh_ack* message to all its active neighbors. A node that receives *reject* informs the waiting neighbors by the *neigh_rej* message.
3. Node i that was waiting neighbors that had proposed in the previous round updates its active neighbors if the reply was *ack*. It waits until it receives the result from all of the proposing neighbors.

At the end of the algorithm, nodes i and j incident to an edge $\{i, j\}$ that is matched have their variable *matched* assigned true value as shown in Algorithm 12.2. Figure 12.3 shows a network of eight nodes with identifiers $1, \ldots, 8$. In the first round, highest identifier nodes 8, 6, 4, and 7 propose randomly to their neighbors. Nodes 3 and 2 select the highest identifier proposals of nodes 8 and 7, these edges are matched, and the nodes incident to these nodes are removed from the graph. In the third round, node 1 gets proposals from neighbor nodes 6 and 4 and accepts the proposal of node 6 as shown in (c). The final matching is shown in (d).

12.2.2.1 Analysis

Theorem 12.1 *Rank_MM algorithm correctly finds a maximal matching of an arbitrary graph in $O(n)$ rounds using $O(nm)$ messages.*

Proof The algorithm continues until a node has a matched edge incident to it (*matched* becomes *true*) or all of its neighbors have matched edges incident to them (*neighs_matched* $= \varnothing$). Therefore, the output is a maximal matching. In the case of a linear network, there will be at least two edge removals in each round, the matched edge and its adjacent edge resulting in $O(n)$ rounds. As there would be a constant

---

Algorithm 12.2 *Rank_MM*

1: **int** i, j ▷ i is this node, j is the sender of a message
2: **set of int** *received, lost_neighs, recvd_prop* $\leftarrow \varnothing$; *curr_neighs* $\leftarrow \Gamma(i)$
3: **message types** *round, propose, unpropose, neigh_ack, ack, reject, inactive, neigh_rej*
4: **boolean** *round_over, round_recvd, prop_flag, reply_flag* $\leftarrow$ *false*
5:
6: **for** $k = 1$ **to** n **do**
7:      {rounds $k, k + 1$ for all nodes}
8:      **if** $(k \bmod 2) = 1$ **then**   ▷ odd numbered round
9:          **while** $\neg$*round_over* **do**
10:             **receive** $msg(j)$
11:                **case** $msg(j).type$ **of**
12:                  $\underline{round(k)}$:   **if** $\neg$*matched* $\wedge$ (*curr_neighs* $\neq \varnothing$) **then**
13:                        **if** $i > max\{curr\_neighs\}$ **then**   ▷ if i am largest
14:                          **select** $j \in curr\_neighs$
15:                          **send** *propose* to j
16:                          **send** *unpropose* to $\{curr\_neighs \setminus \{j\}\}$
17:                          *prop_flag* $\leftarrow$ *true*
18:                        **else send** *unpropose* to *curr_neighs*
19:                        *round_recvd* $\leftarrow$ *true*
20:                  $\underline{propose(k)}$:   *recvd_prop* $\leftarrow$ *recvd_prop* $\cup \{j\}$
21:                        *reply_flag* $\leftarrow$ *true*
22:                  $\underline{unpropose(k)}$:   *received* $\leftarrow$ *received* $\cup \{j\}$
23:               **if** *round_recvd* $\wedge$ (*received* $\cup$ *recvd_prop = curr_neighs*) **then**
24:                  *round_recvd* $\leftarrow$ *false*; *round_over* $\leftarrow$ *true*; *received* $\leftarrow \varnothing$
25:              **end if**
26:          **end while**
27:
28:      **else**   ▷ even numbered round
29:          **while** $\neg$*round_over* **do**
30:             **receive** $msg(j)$
31:                **case** $msg(j).type$ **of**
32:                  $\underline{round(k)}$:   **if** *reply_flag* **then**
33:                        $j \leftarrow max\{recvd\_props\}$
34:                        **send** *ack* to j; **send** *reject* to *recvd_props* $\setminus \{j\}$
35:                        **send** *inactive* to *curr_neighs* $\setminus \{j \cup recvd\_props\}$
36:                        *matched* $\leftarrow$ *true*
37:                      **else if** $\neg$*prop_flag* **then send** *inactive* to *curr_neighs*
38:                      *round_recvd* $\leftarrow$ *true*
39:                  $\underline{ack(k)}$:   *matched* $\leftarrow$ *true*
40:                      **send** *neigh_ack* to *curr_neighs*
41:                  $\underline{reject(k)}$:   **send** *neigh_rej* to *curr_neighs*
42:                  $\underline{neigh\_ack(k)}$:   *lost_neighs* $\leftarrow$ *lost_neighs* $\cup \{j\}$
43:                  $\underline{neigh\_rej(k)}$, $\underline{inactive(k)}$:   ▷ just receive
44:
45:             **if** $msg(j).type \neq round$ **then**
46:                *received* $\leftarrow$ *received* $\cup \{j\}$
47:          **end if**

48:            **if** *round_recvd* ∧ *(received = curr_neighs)* **then**
49:                  *prop_flag, reply_flag* ← *false*; *curr_neighs* ← *curr_neighs* \ *lost_neighs*
50:                  *round_recvd* ← *false*; *round_over* ← *true*; *received, recvd_prop* ← ∅
51:            **end if**
52:        **end while**
53:    **end if**
54: **end for**

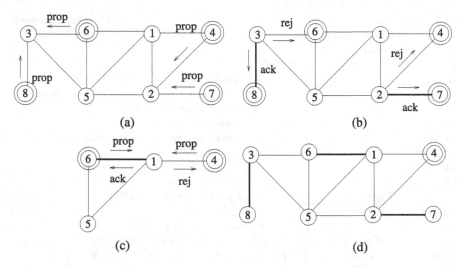

Fig. 12.3 *Rank_MM* algorithm example

times of edge traversals by messages in each round, the total number of messages is $O(nm)$.   □

12.2.3 A Three-Phase Synchronous Distributed Algorithm

As a sample synchronous distributed maximal matching algorithm, we will describe the algorithm due to Panconesi and Rizzi [14] (*PR_MM*), which finds a maximal matching in arbitrary graphs in $O(\Delta + \log^* n)$ rounds. This algorithm has three phases as forest decomposition, three-coloring of forests, and matching described below.

12.2.3.1 Forest Decomposition

In this phase of the algorithm, each node u arbitrarily selects a unique value c_1 from $S = \{1, 2, \ldots, d_u\}$, where d_u is the degree of the node, for edge e_1 incident on it and labels e_1 with this value, which is called the proposal of u for edge e_1. The value c_2 for the next edge e_2 is selected from the remaining values $S \setminus \{c_1\}$. The value of c_1 is sent to the neighbor at the other end of e_1 by the *propose(c_1)* message. The

Algorithm 12.3 *Forest_MM*

1: **set of int** *free_ranks* ← {1, ..., Δ}, *lower_neighs* ← all lower id neighbors, *higher_neighs* ← all higher id neighbors, *edge_ranks* ← ∅, *n_lows* ← |*lower_neighs*|, *childs* ← ∅
2: int parent
3: **for** $k = 1$ to *n_lows* **do**
4:     **select** $r \in$ *free_ranks*, $j \in$ *lower_neighs*
5:     **send** *rank(r)* to j
6:     *free_ranks* ← *free_ranks* \ {r}
7:     *lower_neighs* ← *lower_neighs* \ {j}
8:     *childs* ← *childs* ∪ {j}
9: **end for**
10: **while** *received* ≠ |*higher_neighs*| **do**
11:     receive *rank(r)*
12:     *edge_ranks* ← *edge_ranks* ∪ {$\langle j, r \rangle$}
13:     *parent* ← j
14:     *received* ← *received* ∪ {j}
15: **end while**

value sent by the node with a higher identifier at the ends of edge e_1 is decided as the *rank* of the edge e_1, and e_1 is marked as directed from the lower identifier node to the higher identifier node. This process classifies edges into Δ classes of ranks, after which a forest $F_i(V, E_i) = G = F_1 \cup F_2 \cup \cdots \cup F_\Delta$ of outward rooted arborescences can be constructed, where E_i is the set of edges with ranks i. Based on the foregoing, it can be seen that two edges of the same rank cannot be oriented toward the same node.

We show an adaptation of this phase of *PanRizzi_CSI* in Algorithm 12.3, where node i selects a rank value arbitrarily from free ranks, and node j from the unassigned lower identifier neighbors and sends *rank(j)* message to j.

When the algorithm terminates, all edges are ranked with integers in the range {1, ..., Δ}, and an edge has a direction from the lower identifier node to the higher identifier node, which it stores as its parent. The set of edges with rank k is the forest F_i. It should be noted that labeling the edges of a forest F_i in this manner is not edge coloring and there may be adjacent edges of the same color. Also, there will not be two nodes that have the same parent in an F_i because a node sends distinct rank values to its lower identifier neighbors.

12.2.3.2 Three-Coloring of Forests

In the second phase, the nodes of each rooted arborescence T are colored with three colors in parallel using a suitable algorithm like the one in [5] in $O(\log^* n)$ rounds assuming that each node is aware of its parent and also that the root knows it is the root.

Algorithm 12.4 *PR_MM*

1: Input $G(V, E)$
2: Compute a forest decomposition $F_1, \ldots, F_\Delta$ of G
3: Direct all edges of forests from lower identifier node to higher identifier node
4: Compute a 3-vertex coloring of each F_i in parallel
5: $M \leftarrow \varnothing$
6: **for** $k = 1$ to Δ **do**
7:     **for** $c = 1$ to 3 **do**
8:         **if** $color_i = c$ **then**
9:             **select** e_i incident to i that is outgoing
10:         **end if**
11:         $M_c \leftarrow$ the set of edges selected;
12:         $M \leftarrow M \cup M_c$
13:         **remove** vertices and edges incident to M_c from G
14:     **end for**
15: **end for**

12.2.3.3 Matching

In the last phase of the algorithm, each arborescence that is colored by three colors from the second phase is processed in turn for the three colors c_1, c_2, and c_3. The matching M_1 is formed by choosing an outgoing edge for every node that has color c_1. All the edges incident to the nodes of M_1 are deleted from G as they can not be included in MM. Similarly, M_2 and then M_3 are formed for nodes with colors c_2 and c_3. The maximal matching is finally formed as $M = M_1 \cup M_2 \cup M_3$ as shown in Algorithm 12.4 executed at node i.

A possible execution of *PR_MM* is shown in Fig. 12.4 for the generic network of eight nodes numbered $1, \ldots, 8$. The proposals made by the nodes are shown in (a); (b), (c), (d), and (e) show the forests F_1, F_2, F_3, F_4 that are constructed in the first phase and three colored in the second phase. The final maximal matching obtained in the third phase is shown in (e) with the edges labeled from the forests they are taken from.

Theorem 12.2 *Algorithm* 12.4 *computes a maximal matching in* $O(\log^* n + \Delta)$ *rounds.*

Proof Computing a 3-coloring of the forest concurrently takes $O(\log^* n)$ rounds in the first phase as shown in [14]. Providing the MM takes further $O(\Delta)$ steps resulting in a total of $O(\log^* n + \Delta)$ rounds. $\square$

12.2.4 Matching from Edge Coloring

A recent algorithm proposed by Hirvonen et al. [7] finds maximal matching of a k-edge colored graph G in $O(\Delta)$ rounds, where edges are colored with at most k

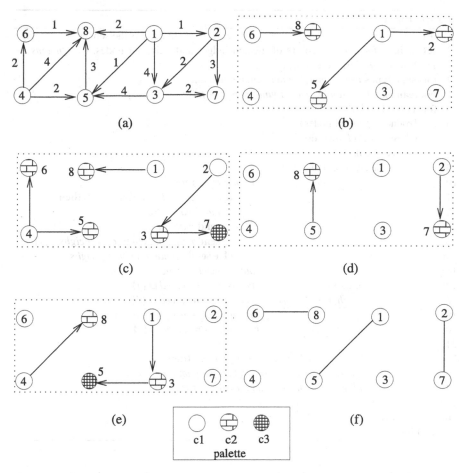

Fig. 12.4 *PR_MM* algorithm example

colors, and a vertex cannot have two edges of the same color incident to it. The idea of the algorithm is to include in matching the edges that are colored with the smallest color c_1 and delete all the edges incident to the end points of edges that are matched from G as the first step. It then proceeds similarly for other colors $c_2, c_3, \ldots, c_k$ in sequence up to k colors.

Algorithm 12.5 (*Ecol_MM*) shows a possible way to obtain a maximal matching from the k-edge colored graph. In each synchronous round starting from round 1, any edge that has a color equal to the number of the round is included in the matching, and any edges that are incident to the vertices of the matched edge are removed from the graph. A node that has been matched sends the *match* message to the neighbor node incident to the other end of the edge colored same as the round number and *neigh_match* to all other neighbors. Any active node that does not have an incident edge with the same color of the round number sends the *unmatch* message to its active neighbors. The colors of incident edges to node i with the neighbor nodes

---

Algorithm 12.5 *Ecol_MM*

```
1:  int i, j                           ▷ i is this node, j is the sender of a message
2:  set of int edge_cols ← colors of incident edges and neighbor nodes; curr_neighs ←
    Γ(i); received, lost_neighs ← ∅
3:  message types round, match, unmatch, neigh_match
4:  boolean matched, round_over, round_recvd ← false
5:  for col = 1 to k do
6:      {round col for all nodes}
7:      while ¬round_over do
8:          receive msg(j)
9:              case msg(j).type of
10:                 round(col):         if ¬matched then
11:                                         if (∃⟨j, c⟩ ∈ edge_cols|c = col) then
12:                                             matched ← true
13:                                             send match to j
14:                                             send neigh_match to curr_neighs \ {j}
15:                                         else send unmatch to curr_neighs
16:                                     round_recvd ← true
17:                 match(col):         received ← received ∪ {j}
18:                 neigh_match(col):   received ← received ∪ {j}
19:                                     lost_neighs ← lost_neighs ∪ {j}
20:                 unmatch(col):       received ← received ∪ {j}
21:
22:         if round_recvd ∧ (received = curr_neighs) then
23:             curr_neighs ← curr_neighs \ lost_neighs; received ← ∅
24:             round_recvd ← false; round_over ← true;
25:         end if
26:     end while
27: end for
```

---

incident to the other end of the edges are stored as tuples $\langle j, c \rangle$ in the set *edge_cols* initially such that edge $\{i, j\}$ incident to nodes i and j has color c.

Figure 12.5 displays the execution of *Ecol_MM* over a sample graph which is 6-colored. Starting from the smallest color 1, edge $\{1, 6\}$ is included in the matching as it was colored with 1, and all the adjacent edges are deleted from the graph. The procedure is repeated for increasing number of colors, where an edge in the subgraph is included, and all its neighbor edges are deleted from graph as these may not be included in the matching. The execution is terminated when a maximal matching with color 4 is reached with edge $\{8, 4\}$. The final matching has edges $\{8, 4\}, \{7, 5\}, \{2, 3\}$, and $\{6, 1\}$ as shown in (d).

12.2.4.1 Analysis

Theorem 12.3 *Algorithm Ecol_MM finds a maximal matching of a graph in $O(k)$ time using $O(km)$ messages.*

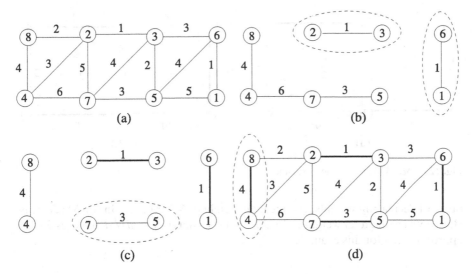

Fig. 12.5 *ECol_MM* execution example

Proof Matching is maximal since the algorithm continues until every node is matched or until all its neighbors are matched. The time complexity of *Ecol_MM* is $O(k)$ as there will be at most k rounds executed. Each edge will be traversed at most once by the *match*, *neigh_match*, or *unmatch* messages in each round. The total number of messages will then be $O(km)$. □

12.3 Weighted Matching

Weighted matching should consider the weights attached to the edges of the graph. In this section, we describe a sequential greedy weighted matching algorithm and a distributed weighted matching protocol.

12.3.1 The Greedy Sequential Algorithm

The greedy algorithm *Seq_MWM* works similar to the sequential matching algorithm, but it selects the heaviest edge e from the graph instead of a random choice and deletes e and all edges incident on the end vertices of e from the graph. This algorithm has the time complexity $O(m \log n)$ [1] with approximation ratio 2. Preis [15] later provided a version of this algorithm that searches for the heaviest edge locally rather than globally and showed that this version has the $O(m)$ time complexity with approximation ratio 2.

Figure 12.6(a) shows an example operation of *Seq_MWM* algorithm in a sample network of six nodes where edges $\{e, d\}$ and $\{a, b\}$ are selected in sequence, in the

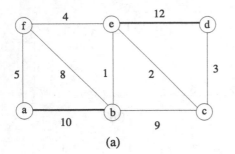 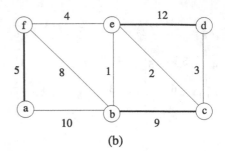

(a)   (b)

Fig. 12.6 *Seq_MWM* execution example

order of decreasing weights, and the total weight of MWM is 22. In (b), MaxWM is shown where the total weight is 26. As can be seen, *Greedy_MWM* achieves 21/26 approximation for this example.

12.3.2 Hoepman's Algorithm

Hoepman [8] provided a distributed version of the Preis Algorithm with $1/2$ approximation to MaxWM. In this algorithm, edge e is included in the MWM if nodes at the ends of e decide that e is the heaviest edge among all their active neighbors. Each node initially has a set S equal to all its neighbors. Node u determines the heaviest edge e_{uv} connected to it and sends a request (*req*) message to the neighbor v at the other end of e_{uv}. If v has e as the locally heaviest edge connected to it, it replies with a request message, and e is added to the MWM. If v has added a different edge to the matching, it drops all its remaining edges from the graph by sending the *drop* message to neighbors, as they will not be included in the MWM.

In Algorithm 12.6 (*Hoepman_MWM*), S is the set of neighbors that are incident to the nondropped edges, and the identifiers of the nodes that requests have been received are kept in the set R. When a drop message is received from a neighbor, node u sends a new request to another neighbor in S.

Figure 12.7 shows an example operation of Algorithm 12.6 in a sample network of six nodes where messages are tagged with the time frame they occur. With initial nodes b and c, nodes e and d send the *req* messages to each other as edges $\{b, c\}$ and $\{e, d\}$ are the heaviest edges for them respectively, and these edges are included in the MWM as shown in (a). The requests of nodes a and f are not granted for edges $\{a, b\}$ and $\{f, e\}$ as these edges are not the heaviest edges for the nodes on the other end of them. Since edges $\{b, c\}$ and $\{e, d\}$ are in MWM, nodes b, c, d, and e send the *drop* messages over the remaining edges as shown in (b). Having received the drop messages, nodes a and b now send the *req* messages to each other over edge $\{a, f\}$, which is the heaviest edge in their current active neighbor set S, which is granted and included in the MWM.

Theorem 12.4 *Hoepman_MWM computes $1/2$ approximation to MaxWM using $O(m)$ messages.*

Algorithm 12.6 *Hoepman_MWM*

1: **set of int** R, S
2: **message types** *req*, *drop*
3: $R \leftarrow \varnothing$
4: $S \leftarrow \Gamma(i)$
5: $c \leftarrow candidate(i, S)$
6: **If** $c = \perp$ **then send** *req* to c
7: **while** $S \neq \varnothing$ **do**
8:      **receive** $msg(j)$
9:      **case** $msg(j).type$ **of**
10:              $\underline{req}$: $R \leftarrow R \cup \{j\}$
11:              $\underline{drop}$: $S \leftarrow S \setminus \{j\}$
12:                          **if** $j = c$ **then** $c \leftarrow candidate(i, S)$
13:                                 **if** $c \neq \perp$ **then send** *req* to c
14:      **if** $c \neq \perp \wedge c \in R$ **then**
15:          **for all** $\in N \setminus \{c\}$ **do**
16:              **send** *drop* to w
17:          **end for**
18:              $S \leftarrow \varnothing$
19:      **end if**
20: **end while**

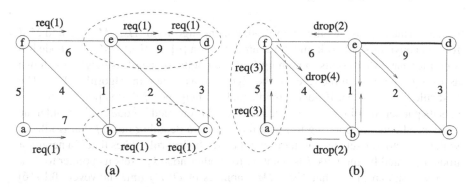

Fig. 12.7 *Hoepman_MWM* execution example

Proof The operation of the *Hoepman_MWM* is similar to *Preis_MWM*, and therefore the approximation ratio is the same. The message complexity is $O(m)$ as each node sends at most one message over its incident edges. A detailed proof of correctness is given in [8].   □

12.4 Self-Stabilizing Matching

We describe three self-stabilizing algorithms for maximal matching in arbitrary graphs in this section. The first algorithm is anonymous and has a central daemon

---

Algorithm 12.7 *Hsu_MM*

1: **R1:**
2: **if** $(i \rightarrow \varnothing) \wedge (\exists j \in N(i))(j \rightarrow i)$ **then**
3:     set $(i \rightarrow j)$
4: **end if**
5: **R2:**
6: **if** $(i \rightarrow null) \wedge (\exists k \in N(i)(\neg k \rightarrow i)) \wedge (\exists j \in N(i))(j \rightarrow null)$ **then**
7:     set $(i \rightarrow j)$
8: **end if**
9: **R3:**
10: **if** $(i \rightarrow j) \wedge (j \rightarrow (k) \wedge k \neq i)$ **then**
11:     set $(i \rightarrow null)$
12: **end if**

---

where the second algorithm uses identifiers of the nodes and has a distributed daemon. The last algorithm is the first linear time self-stabilizing algorithm to find MWM in arbitrary networks.

12.4.1 Hsu and Huang Algorithm

Hsu and Huang provided the first self-stabilizing algorithm (*Hsu_MM*) to find MM for an anonymous network under a central daemon [9]. In *Hsu_MM*, each node has a pointer that either points to nothing (*null*) or points to one of its neighbors shown as $i \rightarrow j$. There are three rules of the algorithm as shown in Algorithm 12.7. The first rule R1 states that if node i is idle and has neighbor j that is pointing to it, it sets its pointer to point to j. The second rule R2 is enabled if node i is idle and none of its neighbors are pointing to it and there exists a neighbor node j that is idle, in which case node i sets its pointer to point to j. The last rule states that if a neighbor node j pointed by the node i is pointing to another node k, i sets its pointer to *null*.

It was shown in [9] that *Hsu_MM* stabilizes in $O(n^3)$ time, however, Tel [16] showed that this algorithm stabilizes in $O(n^2)$ time, and lastly Goddard et al. [6] showed that it actually stabilizes in at most $2m + n$ moves.

12.4.2 Synchronous Matching

Goddard et al. [6] provided a synchronous version of the *Hsu_MM* using a distributed daemon for ad hoc networks and also used identifiers of the nodes. In *Goddard_MM*, each node i has a pointer that either is not pointing to any node, shown by $i \downarrow$, or $i \rightarrow j$ if it is pointing to node j. An edge between nodes i and j is in matching ($\{i, j\} \in MM$) if $(i \rightarrow j \wedge j \rightarrow i)$, in which case $i \leftrightarrow j$. The algorithm evaluates three rules, and if one of them is valid, the node becomes enabled as shown

---

Algorithm 12.8 *Goddard_MM*

1: **R1:**
2: **if** $(i \downarrow) \wedge (\exists j \in N(i) : j \to i)$ **then**
3:     set $(i \to j)$ ▷ accept proposal
4: **end if**
5: **R2:**
6: **if** $(i \downarrow) \wedge (\forall k \in N(i) : k \nrightarrow i) \wedge (\exists j \in N(i) : j \downarrow)$ **then**
7:     set $(i \to min\{j \in N(i) : j \downarrow\})$ ▷ make proposal
8: **end if**
9: **R3:**
10: **if** $(i \to j) \wedge (j \to k$ where $k \neq i)$ **then**
11:     set $i \downarrow$ ▷ back off
12: **end if**

---

in Algorithm 12.8. In Rule R1, node i that is not pointing to a neighbor may choose one of the nodes pointing to it by accepting the proposal made by it. In Rule R2, an idle node i that is not pointing to any neighbor and has at least one neighbor that has a *null* pointer makes a *proposal* by pointing to it if no other neighbor is pointing to it. In case of more than one idle neighbor, it chooses the neighbor with the lowest identifier. In Rule R3, when node i that is pointing to a neighbor j finds that j is pointing to another node $k \neq i$, it changes its pointer to *null* and becomes idle. This situation is called as *backing off*.

12.4.2.1 Analysis

Theorem 12.5 *When Goddard_MM stabilizes, the output matching is maximal.*

Proof Let us assume that the matching produced is not maximal. Every node is either matched or idle as R3 is not enabled since there are no privileged nodes. There will be two idle nodes that are neighbors as the matching is not maximal. However, these two nodes will be privileged to execute as R2 will be enabled for both, a contradiction. It is shown in [4] that this algorithm stabilizes in $n + 1$ rounds.   □

12.4.3 Weighted Matching

Manne et al. [13] proposed the first self-stabilizing algorithm (*Manne_MWM*) that finds the MWM in arbitrary networks. Every node has a pointer m_i to point to the node it will be matched and h_i that shows the weight of the matched edge. Initially, m_i is set to *null*, and h_i is set to 0. Two neighbors i and j are matched when $m_i = j$ and $m_j = i$. The set $S \subset \Gamma(i) : k \in S$ if and only if $k \in \Gamma(i) \wedge w(k, i) \geq h_k$. As such, S includes all neighbors that can be matched with i with a higher or equal weight edge than its currently matched edge. The algorithm consists of a

---

Algorithm 12.9 *Manne_MWM*

1:
2: **procedure** *BestMatch(j)*
3:      **return** $max\{j \in S|w(i, j)\}$
4:
5: **SetMatch**
6: **if** $m_i \neq BestMatch(i) \vee h_i \neq w(i, m_i)$ **then**
7:      $m_i \leftarrow BestMatch(i)$
8:      $h_i \leftarrow w(i, m_i)$
9: **end if**

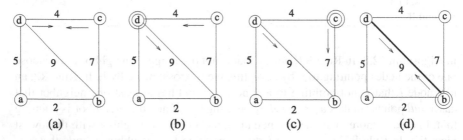

Fig. 12.8 *Manne_MWM* execution example

function *BestMatch* that returns the neighbor $j \in S$ such that $w(i, k)$ is maximal in S and a rule *SetMatch* that updates m_i and h_i according to the value returned by *BestMatch(i)* as shown in Algorithm 12.9.

A possible execution sequence of *Manne_MWM* is shown in Fig. 12.8, where an enabled node is shown by a double circle at each step. Nodes c and d are initially matched. However, c has neighbor b connected with a higher weight edge $\{c, b\}$ than the current one, and d has b as the best match similarly, and therefore both are privileged. The adversarial daemon enables node d, which then points to node b as its best match as shown in (b), and the daemon enables node c, which finds the best match as node b and points to node b in (c), and finally node b executes pointing to d as the best match in (d). It should be noted that node b was enabled initially and could have executed before. The configuration at (d) is stable since none of the nodes will be enabled.

It was shown in [13] that *Manne_MWM* converges in at most $2m + 1$ rounds with $1/2$ approximation under the distributed fair model and $O(3^n)$ time steps under the distributed adversarial model.

12.5 Chapter Notes

Although maximal matching can be achieved in polynomial time, approximation algorithms may be preferred due to their low time and message complexities. We have

Table 12.1 Matching algorithm comparison

	Algorithm	Time Comp.	Msg. Comp.
Unweighted Matching	*Rank_MM*	$O(n)$	$O(nm)$
	Ecol_MM	$O(k)$	$O(km)$
	PancRizzi_MM	$O(\log^* n + \Delta)$	
Weighted Matching	*Hoepman_MWM*	$O(m)$	$O(nm)$
Self-Stab. Matching	*Hsu_MM*	$O(2m + n)$ moves	
	Goddard_MM	$O(n)$ rounds	
	Manne_MWM	$O(2m + 1)$ rounds	

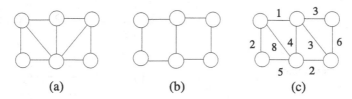

Fig. 12.9 Example graphs for Exercise 1

seen several distributed algorithms to find maximal matching in graphs. Table 12.1 summarizes the algorithms analyzed in this chapter.

Edmonds [3] provided the first polynomial-time sequential algorithm for the maximum weight matching problem. In distributed settings, distributed matching algorithms still remain an active research area. Israeli and Itai [10] presented the first randomized distributed algorithm to compute maximal matching with expected running time $O(\log n)$. Czygrinow et al. [2] presented a deterministic algorithm that computes a 2/3 approximation to MaxM in $O(\log^4 n)$ time. Kuhn et al. [11] showed that any randomized or deterministic distributed algorithm requires $(\Omega \sqrt{(p \log n)/(\log \log n)})$ expected time to compute a 1-approximate maximum cardinality matching. Recently, Lotker et al. [12] gave an algorithm with $(1 - \varepsilon)$-approximation in $O(\log n)$ time for any constant $\varepsilon \geq 0$ for unweighted graphs.

For weighted graphs, Uehera et al. [17] provided a constant time distributed algorithm that finds an MWM of an arbitrary graph with $O(\Delta)$ approximation. Wattenhofer et al. [18] also presented a randomized distributed algorithm with a ratio of 5 to compute an MWM.

12.5.1 Exercises

1. Find examples of MM in (a), MaxM in (b), and MaxWM in (c) of Fig. 12.9.

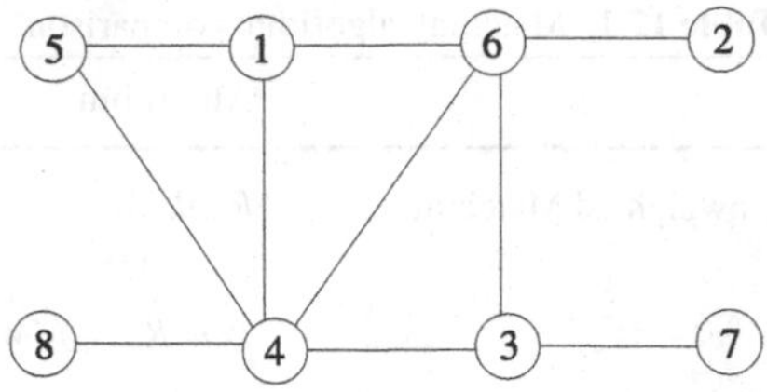

Fig. 12.10 Example graph
for Exercise 2

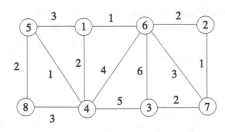

Fig. 12.11 Example graph
for Exercise 3

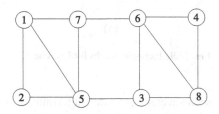

Fig. 12.12 Example graph
for Exercise 4

2. Modify *Rank_MM* algorithm so that the lowest degree node in a neighborhood
   decides on the neighbor node to match in each round. Show also the operation of
   this algorithm in the sample graph of Fig. 12.10.
3. Show the step-by-step execution of the *Ecol_MM* algorithm in the 7-colored
   sample graph of Fig. 12.11, where each edge is labeled by its color.
4. Provide a pseudocode for the greedy distributed weighted matching where a node
   with the highest identifier in a neighborhood selects the heaviest edge incident to
   it and proposes to the neighbor incident to the other end of this edge. Work out
   the time and message complexities of this algorithm. Show also the execution of
   this algorithm in the sample graph of Fig. 12.12.

References

1. Avis D (1983) A survey of heuristics for the weighted matching problem. Networks 13:475–
   493
2. Czygrinow A, Hanckowiak M, Szymanska E (2004) A fast distributed algorithm for approx-
   imating the maximum matching. In: Proc 12th ann European symp on algorithms (ESA),
   pp 252–263
3. Edmonds J (1965) Path, trees, and flowers. Can J Math 17:449–467

4. Goddard W, Hedetniemi ST, Jacobs DP, Srimani PK (2003) Self-stabilizing protocols for maximal matching and maximal independent sets for ad hoc networks. In: Proc international parallel and distributed processing symposium
5. Goldberg A, Plotkin SA (1987) Parallel $(\Delta + 1)$ coloring of constant-degree graphs. Inf Process Lett 4:241–245
6. Hedetniemi ST, Jacobs DP, Srimani PK (2001) Maximal matching stabilizes in time $O(m)$. Inf Process Lett 80:221–223
7. Hirvonen J, Suomela J (2012) Distributed maximal matching: greedy is optimal. In: Kowalski D, Panconesi A (eds) PODC'12. Proc 2012 ACM symposium on principles of distributed computing, Madeira, Portugal, 16–18 July 2012
8. Hoepman J-H (2004) Simple distributed weighted matchings. Technical report, Nijmegen Institute for Computing and Information Sciences (NIII)
9. Hsu S-C, Huang S-T (1992) A self-stabilizing algorithm for maximal matching. Inf Process Lett 43:77–81
10. Israeli A, Itai A (1986) A fast and simple randomized parallel algorithm for maximal matching. Inf Process Lett 22(2):77–80
11. Kuhn F, Moscibroda T, Wattenhofer R (2005) The price of being near-sighted. In: Proc 17th annual ACM-SIAM symposium on discrete algorithms (SODA), pp 980–989
12. Lotker Z, Patt-Shamir B, Pettie S (2008) Improved distributed approximate matching. In: Proc SPAA 2008, pp 129–136
13. Manne F, Mjelde M, Pilard L, Tixeuil S (2007) A new self-stabilizing maximal matching algorithm. In: Proc 14th international colloquium on structural information and communication complexity, pp 96–108
14. Panconesi A, Rizzi R (2001) Some simple distributed algorithms for sparse networks. Distrib Comput 14:97–100
15. Preis R (1999) Linear time 1/2-approximation algorithm for maximum weighted matching in general graphs. In: Meinel C, Tison S (eds) 16th STACS, Trier, Germany, 1999. LNCS, vol 1563. Springer, Berlin, pp 259–269
16. Tel G (1994) Maximal matching stabilizes in quadratic time. Inf Process Lett 49(6):271–272
17. Uehara R, Chen Z (2000) Parallel approximation algorithms for maximum weighted matching in general graphs. Inf Process Lett 76:13–17
18. Wattenhofer M, Wattenhofer R (2004) Distributed weighted matching. In: Guerraoui R (ed) 18th DISC, Amsterdam, The Netherlands, 2004. LNCS, vol 3274. Springer, Berlin, pp 335–348

Chapter 13
Vertex Cover

Abstract A vertex cover of a graph is a subset of its vertices such that at least one endpoint of every edge is incident to a vertex in this set. Vertex cover has many important applications such as facility location and monitoring link failures and placement of routers or agents in a computer network so that every node can be attached to a router or an agent. In this chapter, we will describe sequential and basic distributed algorithms for vertex covering.

13.1 Introduction

The *size* or *cardinality* of a vertex cover is the number of vertices included in the cover. A vertex cover of a graph can be defined formally as follows.

Definition 13.1 (Vertex cover) Given a graph $G(V, E)$, a *vertex cover* $VC \in V$ is the set of vertices such that for any edge $\{u, v\} \in E$, either $u \in VC$ or $v \in VC$. In other words, each edge of G has at least one end point in VC.

A vertex cover is equivalent to an independent set such that given a graph $G(V, E)$, $I \in V$ is an independent set if and only if $V - I$ is a vertex cover. For any $\{u, v\} \notin E$, either $u \notin I$ or $v \notin I$, therefore, $u \in I$ or $v \in I$, which means that $V - I$ covers $\{u, v\}$. To prove the other case, for any two vertices $u \in I$ and $v \in I$, $\{u, v\} \notin E$ as $V - I$ is a vertex cover. Therefore, any two nodes of I are not connected by an edge, meaning that I is an independent set.

For any matching M and any vertex cover VC of G, $|M| \le |VC|$ as each edge can only cover one edge of M. In a bipartite graph G, the requirement for a vertex cover is the same: it should cover all edges of G. In such a graph, the maximum cardinality of a matching is equal to the minimum cardinality of a vertex cover [1].

Definition 13.2 (Minimum vertex cover) A *minimum vertex cover* (*MinVC*) of a graph G is the set of vertices of G that is a vertex cover that has the minimum cardinality among all possible vertex covers.

Definition 13.3 (Minimal vertex cover) The *minimal vertex cover* (*MVC*) is a vertex cover such that removal of a vertex from MVC results in a vertex cover that is not minimal.

K. Erciyes, *Distributed Graph Algorithms for Computer Networks*,
Computer Communications and Networks, DOI 10.1007/978-1-4471-5173-9_13,
© Springer-Verlag London 2013

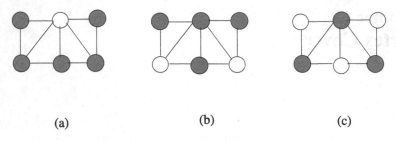

Fig. 13.1 Minimal vertex cover examples

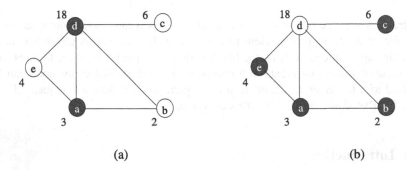

Fig. 13.2 Minimal weighted vertex cover

Definition 13.4 (Minimum weighted vertex cover, minimum connected vertex cover) A *minimum weighted vertex cover* (*MinWVC*) of a weighted graph $G(V, E, w)$, where $E : w \rightarrow \mathbb{R}$, is a set of vertices that is a vertex cover giving a minimum weight among all possible vertex covers. The minimum connected vertex cover (*MinCVC*) is a minimum vertex cover where the graph induced by this cover in G is connected.

Definition 13.5 (Minimal weighted vertex cover, minimal connected weighted vertex cover) A *minimal weighted vertex cover* (*MWVC*) of a graph G is a set V' of its vertices that is a weighted vertex cover where removal of a vertex from V' results in a weighted vertex cover that is not minimal. The minimal connected weighted vertex cover (*MCWVC*) is a minimal weighted vertex cover where the graph induced by this cover in G is connected.

Figure 13.1 displays vertex cover examples where (a) is an MCVC of size 5, (b) is an MCVC of size 4, and (c) is a MinCVC of size 3. A MinVC and a MinWVC are depicted in Fig. 13.2, where weights of vertices are shown next to them. In (a), nodes d and a suffice to cover all edges, and the cover is minimum. The MWVC in (b) includes vertices e, a, b, c and has a total weight of 15, which is lower than the vertex cover of (a). The MCWVC for this graph is a and b.

Finding MinVC is an NP-hard problem, and there are many sequential approximation algorithms that find MVC, MWVC, and MCVC. Examples include using

---

Algorithm 13.1 *Seq1_MVC*

1: Input $G(V, E)$
2: $S \leftarrow E, MVC \leftarrow \varnothing$
3: **while** $S \neq \varnothing$ **do**
4:    **pick** any $\{u, v\} \in S$
5:    $MVC \leftarrow MVC \cup \{u, v\}$
6:    **delete** all edges incident to either u or v from S
7: **end while**

---

depth-first search [9] and semi-definite relaxation [3]. In a distributed setting, however, algorithms that find vertex covers are scarce. Finding a maximal matching and
then including the matched vertices in the cover are, as we will see, a common
method pursued by some researchers. In this chapter, we first show two simple sequential algorithms to find MVC with polynomial execution time. We then describe
distributed algorithms to find unweighted, weighted, and connected vertex covers.
We conclude by two recent self-stabilizing algorithms to find vertex covers.

13.2 Unweighted Vertex Cover

Unweighted vertex cover algorithms assume that the weights of the vertices are the
same, so that they are not considered in the design of these algorithms.

13.2.1 Sequential Algorithms

In this section, we will describe matching-based and degree-based sequential algorithms, which will form the basis of further distributed algorithms for vertex cover.

13.2.1.1 A 2-Approximation Sequential Algorithm

The matching-based sequential algorithm (*Seq1_MVC*) to find the MVC randomly
selects an unassigned edge from the graph and includes the end points of this edge in
MVC and deletes all edges incident to these end points from the edge set as shown
in Algorithm 13.1. Removing an edge from the graph is continued until there are
no edges left, which means that all edges are covered by the included end point
vertices, and therefore the included vertices constitute a vertex cover.

The execution of this algorithm in a graph of six vertices is shown in Fig. 13.3.
The first edge randomly selected is $\{3, 4\}$, and edges $\{2, 4\}$, $\{2, 3\}$, and $\{1, 4\}$ incident
to vertices 3 and 4 are deleted from the graph resulting in the graph of (b). The next
edge selected is $\{1, 5\}$, which results in the inclusion of vertices 1 and 5 in the cover
and covering of all edges. The final vertices of the minimal cover are 1, 5, 4, and 3
as shown in (d), and the size of this vertex cover is 4.

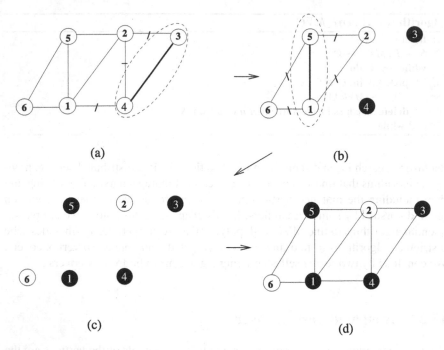

(a)   (b)

(c)   (d)

Fig. 13.3 Random vertex cover example

13.2.1.2 Analysis

Theorem 13.1 *Algorithm* 13.1 *provides a vertex cover in* $O(m)$ *time, and the size of MVC is* $2 |MinVC|$.

Proof Since the algorithm continues until there are no more edges left, every edge is covered, and therefore the output from *Seq1_MVC* is an MVC, taking $O(m)$ time. The set of edges picked by this algorithm is a matching as edges chosen are disjoint, and it is maximal as addition of another edge is not possible. Since two vertices are covered for each matched edge, the approximation ratio for this algorithm is 2.   □

A simple procedure to find an MVC would then consist of finding a maximal matching of the graph G in the first step and including in the maximal vertex cover all the end vertices of the edges found in the first step. The size of a maximal matching of G determines the lower bound of the MVC as the vertex cover cannot have a smaller magnitude than this size; therefore, any minimal vertex cover of G must be at least as large as the size of a maximal matching of G. The following observation can then be made; finding a maximal matching of small size will result in a small size vertex cover when maximal matching is used to find a minimal vertex cover.

Figure 13.4 shows an example network where an MM shown by bold lines in (a) is converted to an MVC in (b) by including the end points of MM in the MVC. It can be seen that the matched vertices 7, 2, and 3 are not needed in the

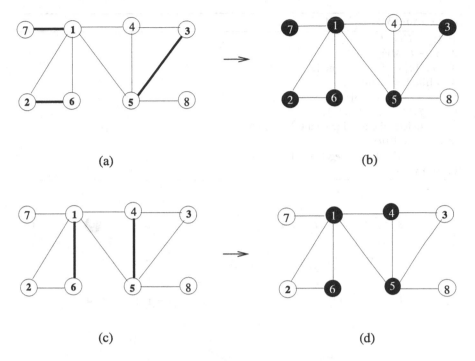

Fig. 13.4 Vertex cover from matching examples

cover. The size of the maximum matching, $|MaxM|$, for this graph is 4 with edges $\{7, 1\}, \{2, 6\}, \{4, 3\}, \{5, 8\}$, and the minimum vertex cover has a cardinality of 4 with vertices 1, 4, 6, 5 or 1, 6, 5, 3. If an MM of size 2, which in fact is the lowest size MM, is selected as in (c), then the MinCVC of (d) is obtained. Selecting $\{1, 6\}, \{5, 3\}$ for MM would also give a MinCVC of size 4 including nodes 1, 6, 5, 3.

13.2.1.3 Highest-Degree-First Algorithm

The aim of this algorithm is to cover as many edges as possible by greedily choosing the highest degree vertex. Instead of choosing a random edge, a vertex that has the highest degree among all the remaining vertices is included in the MVC where symmetries are broken by the magnitude of the identifiers. Also, the degree of a vertex is updated after edges incident to it are deleted from previous iteration as shown in Algorithm 13.2, where the removal of edges continue until there are no edges left.

The execution of this algorithm is shown in Fig. 13.5. The first selected highest degree node is 2, followed by 1 in the second step; and then vertices 6 and 4 in the third and fourth steps, resulting in the vertex cover consisting of vertices 6, 1, 2, and 4 with size 4, which is minimum for this example graph. For different identifiers, ties would have been broken differently, and MVC obtained would be different.

Algorithm 13.2 *Seq2_MVC*

1: Input $G(V, E)$
2: $S \leftarrow E, MVC \leftarrow \varnothing$
3: $degs[n] \leftarrow$ degrees of nodes
4: **while** $S \neq \varnothing$ **do**
5:      $u \leftarrow max\{degs[n]\}$
6:      $MVC \leftarrow MVC \cup \{u\}$
7:      **delete** all edges incident to u from S
8:      $\forall v \in \Gamma(u)$
9:          $degs[v] \leftarrow degs[v] - 1$
10: **end while**

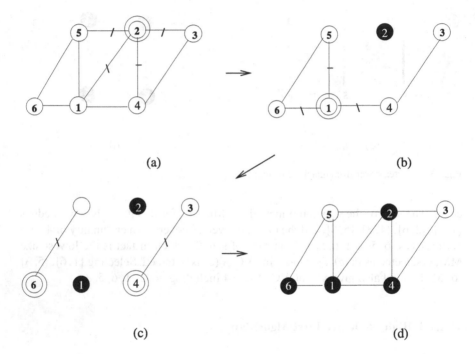

           (a)   (b)

           (c)   (d)

Fig. 13.5 Highest degree vertex cover example

13.2.2 Greedy Distributed MVC Algorithm

The greedy distributed algorithm *Rank_MVC* is similar in operation to the greedy distributed algorithms we have investigated before. This time, however, we choose the node with the current highest degree among its neighbors to be included in the vertex cover as our aim is to cover as many edges as possible. A node that is covered causes the degrees of its uncovered neighbors to be decremented. Initially, all nodes inform their degrees to their neighbors by the *degree*($\langle node\_id, deg \rangle$) messages, and the tuples $\langle node\_id, deg \rangle$ are stored in the set *neigh_degs*. In each round, every node

---

Algorithm 13.3 *Rank_MVC*

1: **set of int** *uncovd_edges* ← Γ(*i*); *neighs_degs, recvd_degs, received, lost_neighs* ← ∅
2: **message types** *round, degree, decide, undecide*
3: **int** *i, j; currdeg* ← |Γ(*i*)|
4: **boolean** *covered, round_recvd, round_over*
5: **send** *degree(d_i)* to Γ(*i*)
6: **while** *recvd_degs* ≠ Γ(*i*) **do**
7:      **receive** *degree(d)*
8:      *neigh_degs* ← *neigh_degs* ∪ {⟨*j, d*⟩}
9:      *recvd_degs* ← *recvd_degs* ∪ {*j*}
10: **end while**
11: **for** *round* = 1 to *n* **do**
12:      {round *k* for all nodes}
13:      **while** ¬*round_over* **do**
14:          **receive** *msg(j)*
15:              **case** *msg(j).type* **of**
16:                  <u>*round(k)*:</u>   **if** ¬*covered* ∧ (*uncovd_edges* ≠ ∅) **then**
17:                                  **if** *currdeg* > *max*{*d*|{⟨*j, d*⟩} ∈ *neigh_degs*} **then**
18:                                      **send** *decide(k)* to *curr_neighs*
19:                                      *covered* ← *true*
20:                                  **else send** *undecide(k, currdeg)* to *curr_neighs*
21:                                  *round_rcvd* ← *true*
22:                  <u>*decide(k)*:</u>   *received* ← *received* ∪ {*j*}
23:                                  *neigh_degs* ← *neigh_degs* \ {⟨*j, d*⟩}
24:                                  *currdeg* ← *currdeg* − 1
25:                                  *uncovd_edges* ← *uncovd_edges* \ {*j*}
26:                                  *lost_neighs* ← *lost_neighs* ∪ {*j*}
27:                  <u>*undecide(k, deg)*:</u>   *received* ← *received* ∪ {*j*}
28:                                  *neigh_degs*(⟨*j, d*⟩) ← *neigh_degs*(⟨*j, deg*⟩)
29:              **if** (*round_recvd*) ∧ (*received* = *curr_neighs*) **then**
30:                  *round_over* ← *true*; *curr_neighs* ← *curr_neighs* \ *lost_neighs*
31:                  *round_recvd* ← *false*; *lost_neighs* ← ∅
32:              **end if**
33:      **end while**
34: **end for**

---

that is not covered and has edges incident to it that are not yet covered checks its *neigh_degs* set. If it finds that it has the highest degree in this set, it decides to be in the vertex cover and informs its neighbors by the *decide* message. Any other neighbor sends the *undecide(deg)* message showing its current degree. Any node that receives a *decide* message deletes the edge that has received this message from the *uncovd_edges* list. It also decrements its current degree as this edge is now covered. This process continues until each node either is covered or has all their edges incident to it deleted from their *uncovd_edges* set as shown in Algorithm 13.3.

Figure 13.6 displays the operation of Algorithm 13.3 in a network of eight nodes. Initially, all nodes exchange their degree information so that they are aware of the degrees of their neighbors. Nodes 3 and 7 have the highest degrees, and they include

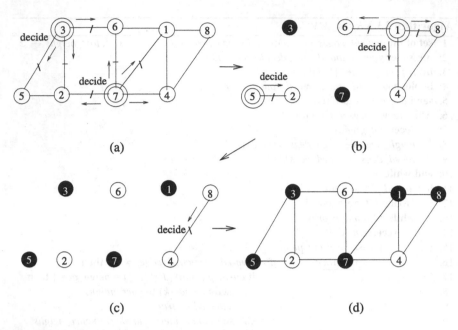

Fig. 13.6 Greedy distributed vertex cover example

themselves in the vertex cover and send the *decide* message to all their neighbors in the first round as shown in (a). The rest of the nodes send the *undecide* messages. The edges incident to these nodes are deleted from the graph, and nodes 5 and 1 are included in the cover in the second round; node 8 is included in the final round as shown (c), and the final vertex cover consisting of nodes 5, 3, 1, 7, and 8 is shown in (d).

13.2.2.1 Analysis

Theorem 13.2 *Algorithm* 13.3 *provides a minimal vertex cover of a graph G in* $O(n)$ *rounds using* $O(nm)$ *messages.*

Proof Similar to the sequential highest degree algorithm, this algorithm continues to remove edges incident to the selected high-degree vertices until there are no more edges left. Therefore, all edges will be covered by the algorithm, and the selected vertices will form a minimal vertex cover. As in the case of a linear network with ordered identifiers, there may only be a single highest degree vertex selected in each round resulting in $O(n)$ rounds in the worst case. Each edge will be traversed by at most one *decide* message in one direction or at most two *undecide* messages in both directions, resulting in $O(m)$ messages per round. The total number of messages used will then be $O(nm)$.   □

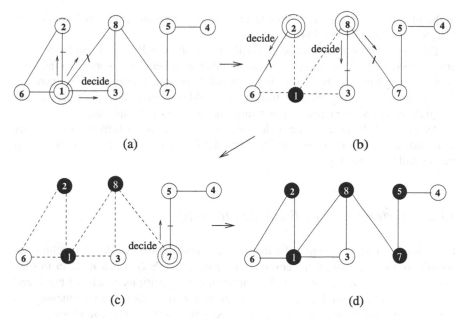

Fig. 13.7 Distributed connected vertex cover example

The greedy algorithm has the same approximation ratio as in the serial case of $O(\log n)$. It is slow as its execution time depends on the number of nodes, and for a large network, this would not be favorable. As was proposed with other distributed greedy algorithms before, the algorithm may finish execution when there are no more edges left by the convergecasting of control messages.

13.2.3 Connected Vertex Cover

We will now attempt to have a distributed algorithm to find a minimal connected vertex cover of a graph where the subgraph induced by the vertex cover is connected. The idea of the algorithm is to first start from the highest degree vertex and iteratively connect highest degree neighbor vertices around it to have a connected vertex cover. In each round of the algorithm, every node checks whether it is the highest degree node in its neighborhood and also whether it has a neighbor node that is already in the vertex cover. If these two conditions are met, the node includes itself in the cover.

The algorithm is similar to the *Rank_MVC*, so we will just show the operation of this algorithm in an example network of Fig. 13.7, where a graph of eight nodes is covered starting from the highest degree node 1. The highest degree nodes in their neighborhood that are connected to 1 are nodes 2 and 8, and they are covered in the second round as shown in (b). The procedure continues with the covering of node 7 in the third round and node 5 in the last round, and the final cover consists of nodes 1, 2, 8, 7, and 5 as shown in (d).

This algorithm starts rather sequentially, but as there will be more nodes covered and so there will be an increase in the number of neighbors adjacent to the connected covered nodes, the probability of parallelism in further rounds will increase. A problem with this algorithm is that as we should start from a unique node that has the highest degree, the degree information should be globally available.

As there will be at least one node added to the cover in each round, the number of rounds is at most n. Similar to the *Rank_MVC*, the total number of messages in transit will be $O(nm)$.

13.2.4 Vertex Cover by Bipartite Matching

Hanckowiak et al. [4] proposed a distributed algorithm that works in synchronous rounds to find a maximal matching of a bipartite graph G. Each node in this algorithm is aware of its partition. Assuming that the partitions consist of black and white nodes, every black node u that is not matched sends a *propose* message to a white node that is free. A white node v receiving multiple *propose* messages decides on one of these and sends the *accept* message to the sender u and the *reject* messages to all other requests. The edge $\{u, v\}$ is included in the matching, and the algorithm continues until black nodes cannot make any more offers, that is, all white nodes are matched to some black nodes.

Polishchuk and Suomela [8] extended the idea of Hanckowiak et al. to anonymous arbitrary graphs where nodes do not have any identifiers by converting such a graph G to a bipartite graph H and then finding a maximal matching in H. Finally, all vertices that are matched are included in the cover. The algorithm proposed is deterministic, constant time and has constant factor approximation ratio achieving 3-approximation in $2\Delta + 1$ rounds.

In this algorithm, called *Bipart_MVC*, each node is replaced with two copies, a black node and a white node. For any existing edge between nodes u and v, the black copy of u is connected to the white copy of v, and a white copy of u is connected to the black copy of v as shown in Fig. 13.8. A maximal matching in this bipartite graph is then computed, and the minimal vertex cover consists of nodes that have their black or white or both copies included in the maximal matching.

The algorithm uses port numbering where all nodes $u \in V$ of $G(V, E)$ assign labels, called *ports*, to their adjacent edges from $1, \ldots, d_u$ arbitrarily. The types of messages used are the *propose* message, which is sent by a node that wants to be matched with a node of opposite color, and the *accept* message, which is sent by a node that receives the *propose* message and decides to be matched with u. Whenever a node sends a message to a port, the neighbor attached to it can receive it in the next round. Nodes propose in odd-numbered rounds and check the *accept* messages in even-numbered rounds.

Each node has two variables *blackpt* and *whitept* to represent its black and white copies and a variable *portnum* that stores the last port that was visited. Every node

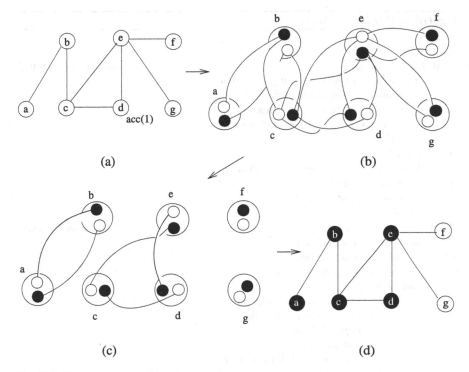

Fig. 13.8 Vertex cover using bipartite matching

that has its black copy unmatched sends the *propose* message to the white copy of
a neighbor in an odd-numbered round in increasing port numbers and waits for the
accept messages in even numbered rounds. Algorithm 13.4 displays a single round
for node i.

As shown in Fig. 13.8, the example network consists of five nodes with identifiers
a, b, c, d, e, f, and g. In Round 1, which is the first odd-numbered round, nodes
send the *propose* messages from their black copies to white copies of the neighbors.
In Round 2, the *accept* messages are received resulting in nodes a, b, d, and e being
matched, which is a vertex cover with the same set of vertices. It should be noted
that, based on the arbitrary port numbering, node b could have proposed to node c
first, and we would have a minimum vertex cover of nodes b, c, d, and e in this case.

13.2.4.1 Analysis

Theorem 13.3 *Algorithm* 13.4 *finds a minimal vertex cover of an anonymous arbitrary graph in* $2\Delta + 1$ *rounds.*

Proof In the worst case, assuming that the highest degree vertex receives a *reject*
message over all of its ports $\{1, \ldots, \Delta\}$ adjacent to it, it will have sent Δ *propose*

Algorithm 13.4 *Bipart_MVC*

```
 1: Input G(V, E)
 2: round(k):
 3: if k mod 2 = 1 then                                          ▷ odd numbered round
 4:     if (blackpt = ⊥) ∧ (1 ≤ count ≤ deg) then
 5:         receive message from port count
 6:         if message.type = accept then
 7:             blackpt ← count, in_MVC ← true
 8:         end if
 9:     end if
10:     if (blackpt = ⊥) ∧ (count ≤ deg) then
11:         count ← count + 1
12:         send propose to port count
13:     end if
14: else                                                         ▷ even numbered round
15:     while received ≠ Γ(i) do
16:         receive message from port j
17:         received ← received ∪ {j}
18:     end while
19:     for all j ∈ Γ(i) in increasing order do
20:         if message(j).type = propose then
21:             if whitept = ⊥ then
22:                 send accept to j
23:                 whitept ← j, in_MVC ← true
24:             else send reject to j
25:             end if
26:         end if
27:     end for
28: end if
```

---

messages, and therefore, we would need Δ odd rounds and $\Delta + 1$ even rounds for a total of $2\Delta + 1$ rounds.   □

Theorem 13.4 *Algorithm* 13.4 *correctly computes a minimal vertex cover.*

Proof We will adapt the proof procedure in [8]. Considering an arbitrary edge $\{u, v\} \in E$, if $blackpt_u \neq \perp$, then $inMVC_u = true$. If this is not the case, then u has sent a *propose* message, and v has responded by a *reject* message meaning that v has already sent an *accept* message to another neighbor and its $whitept_v \neq \perp$ and $inMVC_u = true$. In both cases, either u or v is in MVC, and edge $\{u, v\}$ is covered. The approximation ratio for this algorithm is 3 as shown in [8].   □

13.3 Minimal Weighted Vertex Cover

In a minimal weighted vertex cover, the vertices of graph G have weights, and the aim is to find a vertex cover that has a total minimal weight. We will first show a

---

Algorithm 13.5 *Pricing_MWVC*

1: Input $G(V, E)$
2: $S \leftarrow E, VC \leftarrow \varnothing$
3: **while** $S \neq \varnothing$ **do**
4:     **pick** any $e_{uv} \in S$
5:     **if** $c_u \neq 0 \vee c_v \neq 0$ **then**
6:         $q \leftarrow$ node with $\min\{c_u, c_v\}$
7:         $p_e \leftarrow c_q, q \leftarrow$ *tight*
8:         $VC \leftarrow VC \cup \{q\}$
9:         $S \leftarrow S \setminus \{e\}$
10:     **end if**
11: **end while**

---

sequential and then describe a greedy distributed algorithm for a minimal weighted vertex cover.

13.3.1 Pricing Algorithm

The pricing algorithm (*Pricing_MWVC*) is a sequential algorithm that finds the MWVC of a graph. In this algorithm, each edge should pay a price $p_e \geq 0$ to be covered by vertex u it is incident, and the prices assigned to edges incident to a vertex u cannot exceed the weight of a vertex. Formally,

- an edge $e \in E$ pays a price $p_e \geq 0$ to be covered by a vertex that it is incident.
- for all $u \in V, \sum_{u,v} p_e \leq w_u$.

When the sum of the prices of edges incident to a vertex equals its weight, the vertex is said to be *tight*. A possible implementation of this algorithm is shown in Algorithm 13.5, where each vertex $v \in V$ has a capacity c_i, which is initialized to its weight, and the active edge set S is initialized to E. The algorithm inspects each edge $e_{uv} \in S$, and if u or v has a left capacity, it charges e with the lower of the capacities. When the capacity of u or v becomes 0, it is labeled as a tight node and included in the MWVC.

Figure 13.9 shows the execution steps of *Pricing_MWVC* in an example graph. First, edge $\{d, e\}$ is picked, which can have p_{de} as 2 since vertex d has weight $\{2\}$; d becomes tight and is included in the cover. In (b), the next randomly selected edge is $\{a, e\}$, and the maximum price that can be attributed to this edge is 4, which makes vertex a tight to be included in the cover. Finally in (c), the only edge that can be given a price is $\{b, c\}$, which gets 3 and makes c tight to be included in the MWVC. Nodes a, c, and d, which are in MWVC, are shown with double circles.

At least one new node becomes tight after each iteration of the while loop; therefore, the running time is $O(n)$. When the algorithm terminates, for each edge $\{u, v\}$, either u or v is tight, meaning that the set of tight vertices provides a vertex cover. Pricing algorithm is a 2-approximation algorithm for MWVC [6].

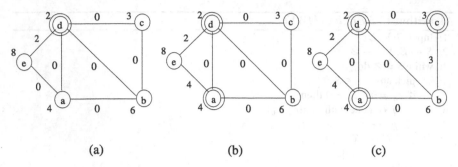

(a)   (b)   (c)

Fig. 13.9 Pricing algorithm example

13.3.2 The Greedy Distributed MWVC Algorithm

The greedy distributed weighted vertex cover algorithm works similarly to the unweighted greedy algorithm with the exception that the weights of the nodes are
considered instead of their degrees. In each round, the lowest weighted node among
active nodes decides to be in the vertex cover and sends a *decide* message to all
active neighbors that still have uncovered edges. We will only show the execution
of this algorithm in Fig. 13.10 in a sample network of eight nodes. All nodes are
labeled by their weights, and ties are broken by the identifiers in the case of equal
weights. Initially, all nodes exchange their weight information so that they can store
the weights of their neighbors. Nodes 1 and 4 are the lowest weighted nodes in their
neighborhoods, and they decide to be in the MWVC. When the edges they cover are
deleted from the graph, the subgraph in (b) is obtained. In the second round, nodes 2
and 3 and finally node 5 are included in the cover, and the MWVC consists of nodes
2, 1, 3, 4, and 5 with a total weight of 12, which is the minimum weighted vertex
cover for this graph, which is also connected. The time and message complexities
for this algorithm are the same as those of *Rank_MVC* algorithm.

13.4 Self-Stabilizing Vertex Cover

As with the general distributed vertex covering of arbitrary graphs, there are not
many self-stabilizing vertex cover algorithms. We will be reviewing an algorithm
that has a less than 2 approximation and an algorithm that forms a bipartite graph of
an anonymous network and then finds a vertex cover by matching.

13.4.1 A $2 - 1/\Delta$ Approximation Algorithm

We have seen that a vertex cover as a result of maximal matching has an approximation of 2. Kiniwa [5] proposed a self-stabilizing algorithm (*Kiniwa_MVC*) with an

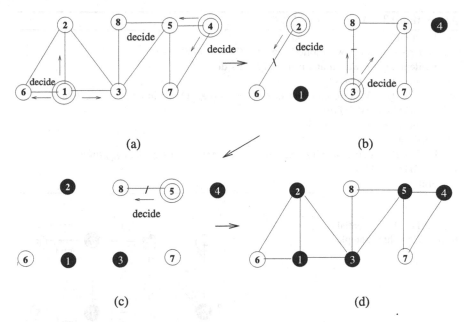

Fig. 13.10 The greedy minimal weight vertex cover algorithm example

approximation ratio of $2 - 1/\Delta$, which is based on a sequential algorithm that considers the degrees of vertices while matching. If the maximal matching is close to the maximum matching, then the vertex cover tends to be large. On the other hand, the greedy approach that selects vertices with high degrees for vertex cover has a ratio of $O(\log n)$ and is a favorable heuristic. Kiniwa integrated these two, namely, maximal matching and highest-degree-first concepts and provided a sequential algorithm to find high-degree-first matching where the vertex with the highest degree is matched first with a neighbor that has the lowest degree. Unlike in the matching-based covering, only the higher-degree vertex is included in the vertex cover.

Initially, all vertices are stored in a sorted list *nodes* from the highest degree vertex to the lowest one. Then, the top vertex u is selected to match with the lowest degree neighbor from *nodes*, u is included in the vertex cover, and both vertices are deleted from *nodes*. This process continues until there are no pairs of vertices left in the list that are neighbors as shown in Algorithm 13.6. A final check is performed to detect any vertices that have been discarded from the list as lower-degree nodes and have lower-degree neighbors that are not in MVC, and such vertices are also included in the MVC.

Figure 13.11 shows the execution of Algorithm 13.7 in a network of eight nodes numbered $1, \ldots, 8$. The highest degree vertex 1 is matched with its lowest degree neighbor vertex 5, and 1 is included in the MVC, but both are deleted from the *nodes* list. Vertex 3 has the next highest degree in the list and is matched with vertex 2 in the second step. In step 3, vertex 7 is matched with 6, and the while loop terminates as the only vertices that remain in the *nodes* list are 4 and 8, which are not neighbors.

Algorithm 13.6 *Kiniwa_Seq_MVC*

1: Input $G(V, E)$
2: *nodes* ← all nodes with sorted degrees, $T ← V$, $MVC ← \varnothing$
3: **while** $\exists u, v \in nodes$: u and v are neighbors **do**
4:     $u ←$ top of *nodes*
5:     $v ← \{x \in V \mid (x \in nodes) \wedge (x \in \Gamma(u)) \wedge (deg(x)$ is minimum among $\Gamma(u))$
6:     $MVC ← MVC \cup \{u\}$
7:     $nodes ← nodes \setminus \{u, v\}$
8: **end while**
9: **if** $(\exists u \notin nodes : u \notin MVC) \wedge (\exists v \in \Gamma(u) : v \notin MVC) \wedge (deg_v < deg_u)$ **then**
10:     $MVC ← MVC \cup \{u\}$
11: **end if**

Fig. 13.11 The sequential Kiniwa algorithm example

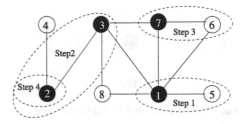

However, vertex 2 is selected next to be included in MVC according to lines 9–10 of the algorithm to give the final MVC shown by black vertices. Kiniwa [5] showed that the approximation ratio for this algorithm is $2 - 1/\Delta$.

The self-stabilizing algorithm is the distributed version of Algorithm 13.6. Each node has a variable, called *color*, which is equal to its degree if unmatched and to the *color* of the proposer if matched. An unmatched node that has neighbors of lower color makes a proposal to the one with the least color. Any node that receives multiple proposals selects the one from the highest color. After node u is matched with another node v, the unmatched nodes that have proposed to u must give up which is provided by total order of degrees. The higher-degree node of a matching is included in the cover as in the sequential algorithm and for any node w that has not proposed; if all neighbors of w have higher degrees than w and are matched, w is not included in the cover, otherwise it is included. The following rules are applied in Algorithm 13.7.

- R1: If node i is not matched with any of its neighbors and if its color is not equal to its degree, its color is set to its degree. This rule corrects colors of unmatched nodes.
- R2: If node i is matched with neighbor j and its color is not equal to the higher-degree value of itself and neighbor j, its color is set to $\max(d_i, d_j)$. This rule corrects colors of matched nodes.
- R3: If node i receives proposals, it accepts the one with the highest degree and sets its color to the degree of the proposer.

Algorithm 13.7 *Kiniwa_MVC*

1: **R1:**
2: **if** $(\forall j \in \Gamma(i) : i \nleftrightarrow j) \wedge (col_i \neq d_i)$ **then**
3:     $col_i \leftarrow d_i$
4: **end if**
5: **R2:**
6: **if** $(\exists j \in \Gamma(i) : i \leftrightarrow j) \wedge (col_i \neq \max(d_i, d_j))$ **then**
7:     $col_i \leftarrow \max(d_i, d_j)$
8: **end if**
9: **R3:**
10: **if** $\exists j_{maxd} : (j_{maxd} \rightarrow i) \wedge (i \nrightarrow j_{maxd})$ **then**
11:     $i \rightarrow j_{maxd}$
12:     $col_i \leftarrow j_{maxd}$
13:     $in\_MVC \leftarrow true$
14: **end if**
15: **R4:**
16: **if** $\exists j \in higher : (i \rightarrow j) \wedge (j \nrightarrow i)$ **then**
17:     $col_i \leftarrow d_i$
18:     $i \rightarrow null$; $in\_MVC \leftarrow false$
19: **end if**
20: **R5:**
21: **if** $\exists j_{maxd} : (j_{maxd} \nrightarrow i) \wedge (\exists k_{mind} : i \nrightarrow k_{mind})$ **then**
22:     $i \rightarrow k$
23:     $in\_MVC \leftarrow true$
24: **end if**
25: **R6:**
26: **if** $(\forall j \in others : (d_i < d_j) \wedge (in\_MVC)) \vee (\exists k \in others : (d_k < d_i) \wedge (\neg in\_MVC))$ **then**
27:     $in\_MVC \leftarrow \neg in\_MVC$
28: **end if**

- R4: A proposal to a higher colored node is discarded.
- R5: If node i is not pointed by any higher-degree neighbors and has lower colored neighbors, it proposes to the minimum degree of such neighbors.
- R6: If node i has all of its neighbors pointing to other nodes and i has the minimum degree, it is not covered. Otherwise, if it is not of minimum degree, it is covered.

Algorithm 13.7 shows a possible way of implementing these rules. The sets *lower*, *higher*, and *others* show neighbors of lower, higher degrees, and neighbors that point (propose) to other nodes other than node i. The variables j_{maxd} and k_{mind} denote the highest degree neighbor j and the lowest degree neighbor k, respectively. The variable matched shows whether a node is matched with another node to cover an edge between them, and node i proposing to node j is shown by $i \rightarrow j$.

It was shown in [5] that this algorithm approximates MinVC by a factor of $(2 - 1/\Delta)$ and stabilizes in $|M_s| + 2$ rounds, where M_s is the matching obtained.

13.4.2 Bipartite Matching-Based Algorithm

A 3-approximation self-stabilizing algorithm that works under a distributed scheduler to find a vertex cover of an anonymous network was proposed in [10]. This algorithm (*Turau_MVC*) is matching based on and follows the idea of [8], where a bipartite graph H is obtained from the original network graph G by replacing each node with a black copy and a white copy. The maximal matching of H is found in the second step, and the nodes that are matched are included in the cover. This time, however, nodes can start from any arbitrary state. The black copy and a white copy of a node are shown by black and white pointers assigned to the port number that their neighbors are attached if there is a match. If these pointers are pointing to the node itself, they are called free pointers. If a node has both pointers as free, it is called a free node. A node is enabled by the two rules shown below:

- R1: This rule is used to set a white pointer of a node. If a node's white pointer is pointing to a node that is not pointing to it by its black pointer, it frees its white pointer. On the contrary, if a node has a black pointer of a neighbor pointing to it and its white pointer is free, it makes its white pointer to point to this neighbor, which means that there is a match.
- R2: This rule is used to control black pointer of a node. If a node's black pointer is free pointing to itself or to a neighbor that has a white pointer pointing to another node, this black pointer is set to another neighbor that has a free white pointer. If there is no such neighbor, the black pointer is freed.

The meanings of *black_matched* and *white_matched* for node u are as follows:

- $black\_matched_u \equiv (blackpt_u \nrightarrow u) \wedge (blackpt_u \rightarrow whitept_v = u)$
- $white\_matched_u \equiv (whitept_u \nrightarrow u) \wedge (whitept_u \rightarrow blackpt_v = u)$

In other words, node u is *black_matched* if its black pointer is pointing to node v that has a white pointer pointing to u. Conversely, node u is *white_matched* if its white pointer is pointing to node v that has a black pointer pointing to u. R1 has priority over R2, so that if both are enabled, a node will execute R1. Algorithm 13.8 shows the implementation of these rules, where *whitept* and *blackpt* are the white and black pointers, respectively. The function *selwhite* attempts to find the neighbor of node u that has a free white pointer, and *selblack* searches a neighbor that has a black pointer pointing to node u.

13.4.2.1 Analysis

Theorem 13.5 *Algorithm* 13.8 *requires* $O(m + n)$ *moves to stabilize under a distributed scheduler, and the obtained vertex cover has 3-approximation.*

Proof Rule 1 can be enabled at most twice per node: first, if it is not pointing to itself, it may be freed, and second, if it accepts an offer from a black pointer. Therefore, there will be a maximum total of $2n$ white moves. Also a node can

---

Algorithm 13.8 *Turau_MVC*

```
 1: procedure selwhite(u)
 2:    find x ∈ Γ(u) : whiteptₓ = x
 3:    return x
 4: end procedure
 5: procedure selblack(u)
 6:    find x ∈ Γ(u) : blackptₓ = u
 7:    return x
 8: end procedure
 9:
10: R1:
11: if (¬white_matched(i) ∧ (whiteptᵢ ≠ i ∨ whiteptᵢ ≠ selblack(i) then
12:    if whiteptᵢ ≠ i) then
13:       whiteptᵢ ← i
14:    else
15:       whiteptᵢ ← selblack(i)
16:    end if
17: end if
18: R2:
19: if ¬black_matched(i) ∧ (blackptᵢ = i ∨ blackptᵢ → whiteptⱼ ≠ blackptᵢ) ∧ (blackptᵢ ≠
       selwhite(i)) then
20:    blackptᵢ ← selwhite(i)
21: end if
```

---

make at most once black move using its black pointer. Therefore, there can be at most $2d(i)$ black moves for a node. The total number of moves will then be $2n + 2\sum_{i \in V} d(i) = 2n + 4m$, hence $O(m + n)$ moves.   □

This algorithm, which is similar to the Polishchuck algorithm, has the advantage of reading the pointer variables of neighbor nodes as it is self-stabilizing and hence can decide whether a neighbor node is pointing to another node.

13.5 Chapter Notes

Finding a minimal vertex cover of a graph remains a fundamental problem in a distributed setting due to the many applications it has. Algorithms with constant approximation ratio that are not dependent on the number of nodes are favorable. We have seen in this chapter that computing a matching first and then converting this matching to a vertex cover method are pursued by many researchers. In this respect, distributed maximal matching algorithms such as Hanckowiak et al. [4], which works for $O(\log^4 n)$ rounds, or by Panconesi and Rizzi, which requires $O(\Delta + \log^* n)$ rounds [7], or self-stabilizing matching algorithms due to Hsu et al. and Goddard et al., which we have seen in Chap. 11, can be used as the first step.

The algorithms that find MVC directly are scarce, and this area may be a potential research topic for researchers. The Connected Vertex Cover Problem was first

Fig. 13.12 Example graph
for Exercise 2

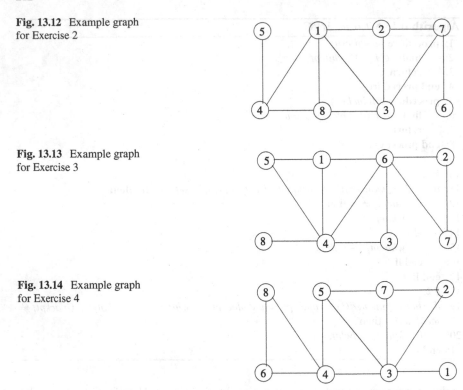

Fig. 13.13 Example graph
for Exercise 3

Fig. 13.14 Example graph
for Exercise 4

defined in 1977 by Garey and Johnson [2], who showed it to be NP-hard even when
restricted to planar graphs with maximum degree 4. The first constant ratio algorithm for MVC was given by Carla Savage [9], showing that the internal nodes of
any depth-first search tree provide a 2-approximation for vertex cover. These nodes
induce a connected subgraph, and since the size of MCVC is greater than or equal
to MVC, the approximation ratio is 2.

13.5.1 Exercises

1. Find MVC by showing the step-by-step execution of *Seq2_MVC* in the graph of
   Fig. 13.4.
2. Provide the FSM-based version of *Rank_MCVC* algorithm described in
   Sect. 13.2.3, which finds the MCVC by greedily selecting the remaining highest
   degree nodes. Provide its FSM diagram, a pseudocode and work out the time and
   message complexities. Show also step-by-step execution of this algorithm in the
   example graph of Fig. 13.12.
3. Show the execution of MWVC algorithm in the sample graph of Fig. 13.13,
   where weights are shown as labels of vertices.

4. Provide the pseudocode of a greedy distributed minimal weighted connected vertex cover based algorithm using the algorithm described in Sect. 13.3.2. Work out its time and message complexities. Also, show the operation of this algorithm step-by-step in the example graph of Fig. 13.14 where each vertex is labeled by its weight.
5. Provide the DFS tree of the graph of Fig. 13.14 starting from the vertex with the heaviest weight. Show that all nonleaf nodes of this tree constitute a vertex cover.

References

1. Babu AC, Ramakrishnan PV (1991) New proofs of Konig–Egervary theorem and maximal flow–minimal cut capacity theorem using o.r. techniques. Indian J Pure Appl Math 22(11):905–911
2. Garey MR, Johnson DS (1977) The rectilinear Steiner tree problem is NP complete. SIAM J Appl Math 32:826–834
3. Halperin E (2002) Improved approximation algorithms for the vertex cover problem in graphs and hypergraphs. SIAM J Comput 31(5):1608–1625
4. Hanckowiak M, Karonski M, Panconesi A (1998) On the distributed complexity of computing maximal matchings. In: Proc 9th annual ACM-SIAM symp on discrete algorithms. SIAM, Philadelphia, pp 219–225
5. Kiniwa J (2005) Approximation of self-stabilizing vertex cover less than 2. In: Herman T, Tixeuil S (eds) Self-stabilizing systems. LNCS, vol 3764. Springer, Berlin, pp 171–182
6. Kleinberg J, Tardos E (2005) Algorithm design. Addison-Wesley, Reading. Chapter 11
7. Panconesi A, Rizzi R (2001) Some simple distributed algorithms for sparse networks. Distrib Comput 14(2):97–100
8. Polishchuk V, Suomela J (2009) A simple local 3-approximation algorithm for vertex cover. Inf Process Lett 109(12):642–645
9. Savage CD (1982) Depth-first search and the vertex cover problem. Inf Process Lett 14(5):233–237
10. Turau V, Hauck B (2011) A fault-containing self-stabilizing $3 - 2/(\Delta + 1)$-approximation algorithm for vertex cover in anonymous networks. Theor Comput Sci 412:4361–4371

Part III
Ad Hoc Wireless Networks

Chapter 14
Introduction

Abstract Ad hoc wireless networks communicate over wireless links without a fixed communication structure. As they can be deployed easily and quickly, these networks have many applications such as tactical operations, monitoring, military networks, and rescue operations. In this chapter, we investigate the basic principles, design issues, and modeling and simulation of important types of ad hoc wireless networks.

14.1 Ad Hoc Wireless Networks

The computing nodes of a *wireless network* are equipped with wireless communication and networking facilities. Wireless networks can either be constructed as *infrastructured* or *ad hoc*. In infrastructured networks, a stationary wired backbone, which consist of routers, access points, and servers, is used to provide communication between the nodes of the network as shown in Fig. 14.1, where five mobile hosts (MHs) access the backbone that consists of three routers and a server. In such a network, mobile or stationary nodes do not communicate directly. In case of a disaster, the backbone may not function causing total loss of communication among the hosts, as experienced in some disasters. Cellular networks are one type of infrastructured wireless networks.

On the contrary, an *ad hoc wireless network* does not have any fixed structure, and each node participates in communication by forwarding messages as shown in Fig. 14.2. An ad hoc wireless network may consist of heterogeneous nodes with different capabilities or homogeneous nodes that have same wireless communication facilities. Two types of wireless ad hoc networks that are in common use are the *mobile ad hoc networks* and the *wireless sensor networks*, which we will investigate in the following sections.

14.2 Mobile Ad Hoc Networks

A *mobile ad hoc network* (*MANET*) consists of autonomous mobile nodes that can move freely and randomly and communicate without any communication infrastructure. Examples of MANETs include emergency (search and rescue) operations,

K. Erciyes, *Distributed Graph Algorithms for Computer Networks*,
Computer Communications and Networks, DOI 10.1007/978-1-4471-5173-9_14,
© Springer-Verlag London 2013

Fig. 14.1 An infrastructured network

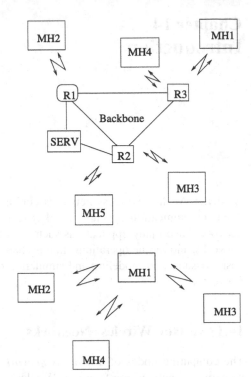

Fig. 14.2 An ad hoc wireless network

disaster relief operations, inter-vehicle networks, and military networks. Each node of a MANET is autonomous and acts both as a host and a router to transfer messages. If two nodes of a MANET are within radio transmission ranges of each other, they communicate directly. Otherwise, they can use *multi-hop communication*, where intermediate nodes are used to transfer a message between the two nodes. As there is no infrastructure and central control, the management of the network is achieved by the individual nodes. The following are considered as the main challenges in a MANET:

- *Routing*: Multi-hop routing is usually the only choice in a MANET. Also, as the topology is dynamic, routing tables need to be updated frequently. Figure 14.3(a) displays an example where nodes u and v moving close to each other form a communication link when they detect each other by beacon messages. In (b), a third node w comes to the vicinity of node u. At this point, node u broadcasts its new neighbor w that is received by node v. As node v has not detected node w, it determines that it cannot reach node w directly but can do this via node u, so it updates its routing table. In (c), however, the situation changes as the nodes are moving, and now node u can reach node v via node w, and the route tables are again updated. Routes in a mobile network may be updated periodically, so that there is always a route between any sender–receiver pair, in which case the routing algorithms are called *proactive*. In *reactive* routing protocols, routes are only formed when needed. Both methods have advantages and disadvantages as

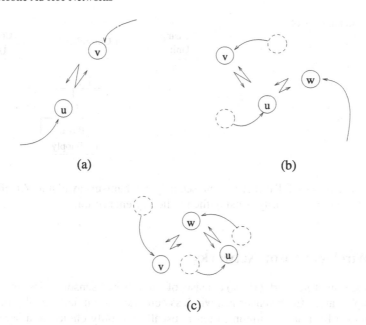

(a)   (b)

(c)

Fig. 14.3 A MANET example

described in Chap. 16. Routing via a dynamic backbone using a connected dominating set is a widely used method and will be discussed in Chap. 15.

- *Topology Control*: Topology control is difficult in a MANET where connections
change rapidly in an unpredictable manner. Constructing a graph with sparser
communication links is a common method for topology control that we will investigate in Chap. 15. Clustering of nodes provides an efficient method to manage
resources and also routing in a MANET and is also described in Chap. 15.

- *Channel Access*: The wireless channels that the nodes communicate are subject
to noise, fading, and interference conditions due to multiple access. When two
or more nodes are within the transmission ranges of each other, their simultaneous transmissions may interfere, causing collisions of packets. A *hidden terminal*
cannot sense a transmitting neighbor or correctly receive a reservation packet
from its receiver and therefore may start transmission that may collide with an
ongoing one. In fact, the hidden terminal problem is the main source of collisions in MANETs. Medium access control protocols should provide mechanisms
to reduce collisions. Using separate channels for control and data packets and
sometimes separate channels for different nodes is a common method to reduce
collisions. Synchronization-based methods provide dedicated time slots for nodes
to prevent collisions.

- *Quality of Service (QoS)*: Provision of QoS in terms of bounded delay and minimum bandwidth especially for multimedia applications is difficult in a MANET
due to the mobility of the nodes and channel access problems outlined.

- *Security*: The shared wireless medium is open to legitimate users and malicious
attackers. Lack of secure routers under a central controller means that attackers

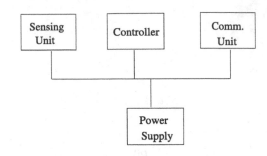

can disable a MANET if there are no security mechanisms available. An effective
way of achieving security is using the public key encryption.

14.3 Wireless Sensor Networks

A wireless sensor network (*WSN*) consists of a number of sensor nodes spread over
a geographic area. Each sensor has wireless communication and signal processing
capabilities. Multi-hop communication is usually the only choice in a WSN, and
a central node with more advanced communication capabilities called the *sink* or
fusion center gathers all the data from the sensors, possibly performs further processing of this data, and transfers it to another computer for advanced data processing. Figure 14.4 shows the architecture of a sensor node that has a controlling unit,
which is usually a microcontroller, a sensing unit which senses a physical event and
provides the data representing the signal to the controller, and the communication
interface which transmits and receives data in the form of radio signals.

In a sensor network, many-to-one communication is performed, where data from
simple sensors with short-range communication capabilities are transferred to the
sink or fusion center for further processing as shown in Fig. 14.5, where nodes
send their sensed data to the nearest sensor neighbor toward the sink over a spanning tree rooted at the sink. A sensor node that receives data from neighbors needs
to eliminate redundant data before sending it to another neighbor, which is called
data aggregation. For example, only the average temperature value received from a
number of sensors in a particular region may be transmitted toward the root by an
intermediate node to reduce the number of messages and therefore save energy.

WSN designers and implementors are faced with the following challenges.

- *Scalability*: A sensor network may consist of hundreds or thousands of nodes.
  Protocols such as for routing and algorithms for these networks should be scalable.
- *Fault Tolerance*: Failure of a single node should not affect the overall operation
  of a sensor network, and connectivity of the network should be maintained using
  new links if required.
- *Node Deployment*: Nodes can be deployed randomly or regularly with the aim of
  providing maximum coverage of the area to be monitored by the sensors. Also,
  the deployment should provide a connected graph of the network.

Fig. 14.5 A wireless sensor network

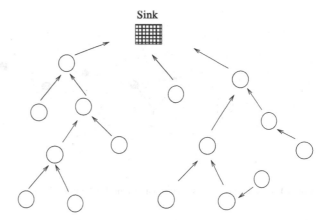

Sink

- *Power Management*: Nodes of a sensor network rely on batteries with limited life-time, and replacing these batteries may be difficult or impossible in many applications where sensors are deployed in harsh conditions. Transmission and reception of messages consume much more energy than local computations, therefore, protocols and algorithms should be designed to work with minimum number of messages. An efficient method to conserve energy is to put some of the sensors into sleep mode when there are not many activities to be sensed. An important requirement in these methods would then be to provide full coverage of the area by the awake sensors.

14.4 Ad Hoc Wireless Network Models

Modeling and simulation are frequently used to evaluate the performance of wireless systems. A model is a simplified representation of a system that aids the understanding and investigation of the real system. In this section, we review basic communication models that consider the inherently broadcast communication property of the wireless networks as in [4].

14.4.1 Unit Disk Graph Model

A common model for ad hoc wireless networks that considers the broadcast transmission of the nodes is the *Unit Disk Graph Model* defined as follows.

Definition 14.1 (Unit disk graph) A *Unit Disk Graph* (*UDG*) is a graph where each node $u \in V$ of the graph is identified by a radius of magnitude 1 and edge $e \in E$ exists between the two vertices u and v if and only if the Euclidian distance between them is ≤ 1.

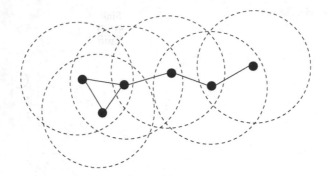

Fig. 14.6   A network constructed using Unit Disk Graph Model

In other words, two nodes u and v in the network are neighbors if the Euclidean distance between them is at most unity, which is the normalized transmission radius that is the same for every node [3]. Figure 14.6 displays a network constructed using the UDG model, where transmission range of nodes are shown by circles with the nodes at the centers.

The UDG model has two major drawbacks. First, it assumes nodes with identical transmission capabilities, which may not be valid either due to heterogenous nodes or because transmission of the nodes may be disturbed by obstacles. Second, it does not consider node weights, which are very useful to specify other parameters such as the mobility and residual energy of nodes. However, the UDG model is still widely used for modeling ad hoc wireless networks due to its simplicity. There are few variations of UDG for topology control, which we will investigate in Chap. 15.

14.4.2 Quasi Unit Disk Graph Model

A *Quasi Unit Disk Graph (QUDG)* is an extended UDG model for ad hoc networks where each node is identified by two disks. The inner disk has a radius r, and the outer disk has a radius 1 as in the UDG. The edges between a node u and its neighbor v at a distance d is determined as follows:

- An edge between node u and v exists if $d_{uv} \leq r$.
- There may be an edge between nodes u and v if $r < d_{uv} \leq 1$.
- There are no edges between node u and v if $d_{uv} > 1$.

Figure 14.7 shows the edges of node u with neighbors v, w, s. The edge $\{u, v\}$ is a solid edge, edge $\{u, w\}$ is a possible edge, and there is no direct communication possibility between nodes u and s. The effect of obstacles and heterogenous transmission capabilities in a network can be handled by adjusting the parameter r, however, the node weight problem remains.

Fig. 14.7 A network
constructed using Quasi Unit
Disk Graph Model

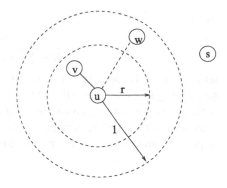

14.4.3 Interference Models

Interference in an ad hoc wireless network is caused by the simultaneous transmission of nodes that are within transmission ranges of each other. Once interference occurs, the interfered message has to be retransmitted. Topology control algorithms usually provide a sparser graph with an average low degree, reducing the number of neighbors that a node has, hence reducing interference. Rather than this indirect approach, recent approaches attempt to solve the interference problem explicitly [6].

Topology control algorithms to prevent interference explicitly may be broadly classified as *sender centric* and *receiver centric*. In sender centric algorithms, the number of nodes affected by the communication link is explored. These algorithms generate a subgraph by eliminating the edges that have high coverage. The resulting subgraph must be connected and have spanner properties. The *coverage* of an undirected edge (u, v) is the number of nodes covered by disks of both u and v in the UDG that are the nodes affected when u and v are communicating. One approach is to label the edges of the graph with the total number of nodes covered by both of its end points and to include in the final subgraph the edges in increasing order until all nodes are covered as in LIFE [9].

In receiver centric algorithms, however, the aim is to find the number of nodes that are affected other than the receiving node. The *interference* of node v is the number of nodes other than v that are affected by message reception at v. In this sense, the interference of node v is the number of disks that include v in the UDG. Receiver centric algorithms attempt to find the interferences for each node in contrast to the computation of the coverage of links in sender centric algorithms and use this parameter to obtain a connected sparser subgraph as in the Nearest Component Connector (NCC) method described in [9].

A message transmitted to node v will be received if the signal strength is sufficiently high. However, the message can be corrupted because of the interference even if the signal strength is high enough. For the message to be received correctly, the signal strength should be stronger than the ambient noise and other interfering signals. The *Signal-to-Interference-and-Noise Ratio* (*SINR*) inequality is given by [10]:

$$\frac{s(u, v)}{N + \sum_{w \in H} s(w, v)} \geq \gamma, \tag{14.1}$$

where $s(u, v)$ is the strength of the signal for the message sent between nodes u and v; $N \geq 0$ is the noise induced by the environment and the circuits of node v; H is the set of nodes that transmit in parallel to u, and γ is a constant related to hardware and software characteristics of the receiving node v. SINR-based approaches to prevent interference try to find low-interference topologies based on SINR.

14.5 Energy Considerations

Energy consumption of a node is important especially for sensor networks where changing of batteries may not be possible. In a sensor network, energy is spent by sending messages and listening to the medium, the latter being dominant in general. Therefore, we need *power saving* techniques that will disable nodes for a while in order to conserve energy.

Energy management can be achieved by assigning states to the nodes of a sensor network as *transmit*, *receive*, *idle*, and *sleep* modes. A node is at *transmit* state while transmitting a packet and *receive* state while receiving a packet. When a node is neither transmitting or receiving data, it is at *idle* state. A node consumes power at *idle* state as it has to listen to the medium so that it can enter *receive* state if it detects a transmission. In *sleep* state, a node cannot transmit or receive any data, therefore, it has to be awaken by an internal event such as an interrupt to be able to switch to *idle* state to detect any transmissions. The *sleep* state requires the lowest energy. The choice of nodes that will sleep may be performed according to some schedule with the aim of maintaining connectivity of the network graph at all times. Such schedules may be activated by topology control algorithms at network layer or by performing time division multiplexing at MAC layer where each node is given a time slice to be active.

Node u that communicates with a neighbor transmits a signal with power P_t, and this signal is received at the receiving node v with power P_r, which is smaller than P_t since the signal gets weaker as the Euclidean distance d_{uv} between the two nodes gets larger. P_r is given by

$$P_r(d) = a \cdot \frac{P_t}{d^x}, \tag{14.2}$$

where a is a constant related to physical transmission characteristics such as antenna and wavelength, x depends on the medium and is usually between 2 and 4. The transmitting node u can adjust its transmitting power for the message to be received with adequate power at the receiving node v. The energy consumption of a distributed algorithm can then be computed by assigning energy levels to the edges in the network graph and summing all these communication costs over all the transferred messages.

14.6 Mobility Models

Mobility models simulate the movements of nodes in mobile and sensor networks and are mainly used to evaluate the performance of the protocols designed in the simulators. In the *entity models*, each node is assumed to move independently, and in the group mobility *models*, nodes are clustered around groups, and the motion of the groups is modeled. The most common entity models are *Random Waypoint*, *Random Direction*, and *Gauss–Markov* mobility models [1, 4].

- *Random Waypoint Model*: The two important parameters of this model are the velocity and pause time of each node. Every node moves independently at constant velocity toward a randomly designated destination. When a node arrives at the destination, it waits for a designated time, and if the pause time is zero, the node has a continuous mobility. A small velocity and a long pause time provide a stable and slowly changing topology, whereas the contrary means a dynamic topology.
- *Random Direction Model*: In Random Direction Model, the direction rather than the destination is randomly chosen. A node travels in the selected direction until it reaches a boundary, waits there for a designated time, and then chooses another direction.
- *Gauss–Markov Model*: In this model, the velocity of a mobile node is determined using Gauss–Markov stochastic processes where the velocity at time slot t depends on the velocity in the previous slot, at time $t-1$. Different mobility models are provided by tuning the parameters and changing the randomness of Gauss–Markov processes.

14.7 Simulation

Simulation is the manipulation of the model of a system enabling to observe the behavior of the system in a setting similar to real life. As the number of nodes of a MANET and a WSN is increased, experimentation and evaluation of a distributed algorithm or a protocol for these networks become extremely difficult. For this reason, simulators that provide the virtual platform for the distributed application are widely used. In this section, we will briefly review some simulators that are widely used for ad hoc wireless networks.

14.7.1 ns2

*ns*2 (Network Simulator 2, version 2.34) [2] is a discrete event simulator that provides simulation of TCP, routing, and multicast protocols over wired and wireless networks. It is one of the most widely used simulators for MANETs and WSNs. In order to simulate a program using *ns*2, the user first writes the code in C++, and the type of simulation is specified in OTcL script. As the final step, the protocol is executed, and outputs in the form of trace files or direct protocol outputs are collected to be analyzed.

Routing of messages is handled by five alternative ad hoc routing protocols: Destination Sequence Distance Vector (DSDV), Dynamic Source Routing (DSR), Temporally Ordered Routing Algorithm (TORA), Ad hoc On-demand Distance Vector (AODV), and Protocol for Unified Multicasting Through Announcements (PUMA). We will be investigating some of these protocols in Chap. 16.

14.7.2 TOSSIM

TinyOS is a sensor network operating system that runs on sensor nodes, called *motes*, and TOSSIM is a discrete event simulator for TinyOS [5]. The TinyOS has a component-based programming model and uses a language called nesC with a C-like syntax. A TinyOS program consists of computational entities, called *components*, that communicate with other components using commands and events. A task is the basic execution unit in TinyOS.

The TOSSIM architecture consists of compilation support for simulation, a discrete event queue, a small number of hardware abstraction components, mechanisms for extensible radio and Analog-to-Digital Converter (ADC) models, and communication services [5]. TOSSIM models hardware interrupts by an event queue and emulates hardware components such as ADC, clock, and EEPROM. TOSSIM provides two built-in radio models: simple radio model, which simulates error-free transmission, and lossy radio model, which simulates packet loss in a probabilistic manner.

14.7.3 Other Simulators

OMNeT++ is a modular object-oriented discrete event network simulator that can perform simulations of traffic modeling of telecommunication networks, protocol modeling, and modeling of MANETs [7]. An OMNeT++ model consists of hierarchically nested modules written in C++ and communicating by messages. The modules are managed with a high-level script language called NED.

GloMoSim is a scalable parallel discrete-event simulator for wired and wireless networks. GloMoSim has a layered architecture, and the protocols are performed by Parsec, which is a C-based simulation language [11].

Sinalgo is a network simulator for testing and validating network protocols and algorithms. It can provide quick prototyping of the network algorithms for over 10 K nodes, and it also has 2D and 3D support, asynchronous and synchronous simulation, and UDG and QUDG models. Sinalgo provides visualization of the network graph [8].

14.8 Chapter Notes

We have described mobile ad hoc and sensor networks and listed important problems encountered in the design and realization of these networks. In terms of interacting with the external world, MANETs provide services to the users, whereas WSNs, in general, interact with the environment and send their sensed data to the sink/fusion center for further processing. Mobility management is of primary concern in a MANET, whereas designing power efficient communication strategies is imperative in WSNs. Designing efficient routing algorithms is important in both networks with somehow different goals. In a MANET, primary goal is to manage mobility either by frequent route table updates or discovery of the routes only when needed. In a sensor network, however, a query from the sink may initiate data transfer, and the location of the nodes may be considered to deliver messages to the sink efficiently.

Simulators provide realistic evaluation and verification of algorithms and protocols designed for ad hoc wireless networks. Other than the described simulators, we will provide a detailed discussion of a POSIX thread-based simulator called *ASSIST* that we have designed and implemented in Chap. 18.

In the remaining chapters of this part, we restrict our investigation in these networks to efficient distributed graph algorithms to solve problems related to topology control, namely, constructing local graphs, backbone construction and clustering, and also routing in MANETs and WSNs.

14.8.1 Exercises

1. Discuss briefly the main differences between a MANET and a WSN in terms of structure and energy constraints.
2. Compare sender centric and receiver centric interference models by stating their advantages and disadvantages.
3. Give an example of data aggregation in a WSN and provide a pseudocode of an algorithm that will perform this operation.
4. Compare the UDG and QUDG models. Comment on the choice of the inner radius value for the QUDG.

References

1. Bai F, Helmy A (2004) A survey of mobility modeling and analysis in wireless ad hoc networks. In: Wireless ad hoc and sensor networks. Kluwer Academic, Dordrecht
2. Bajaj S et al (1999) Improving simulation for network research. Technical report 99-702, University of Southern California
3. Clark BN, Colbourn CJ, Johnson DS (1990) Unit disk graphs. Discrete Math 86:165–177

4. Erciyes K, Dagdeviren O, Cokuslu D, Yilmaz O, Gumus H (2011) Modeling and simulation of mobile ad hoc networks. In: Mobile ad hoc networks: current status and future trends. CRC Press/Taylor & Francis, Boca Raton/London
5. Levis P, Lee N, Welsh M, Culler D (2003) TOSSIM: accurate and scalable simulation of entire TinyOS applications. In: Proc of the 1st international conf on embedded networked sensor systems, Los Angeles, California, USA
6. Locher T, von Rickenbach P, Wattenhofer R (2008) Sensor networks continue to puzzle: selected open problems. In: Proc of the 9th international conf on distributed computing and networking, Kolkata, India
7. OMNET++. Available: http://omnetpp.org/ (2010, 01.08.2010)
8. Sinalgo—simulator for network algorithms. Available: http://disco.ethz.ch/projects/sinalgo/ (2010, 01.08.2010)
9. von Rickenbach P, Wattenhofer R, Zollinger A (2009) Algorithmic models of interference in wireless ad hoc and sensor networks. IEEE/ACM Trans Netw 17:172–185
10. Wagner D, Wattenhofer R (eds) (2007) Algorithms for sensor and ad hoc networks. Springer lecture series. Springer, Berlin. Section 6.2.1
11. Zeng X, Bagrodia R, Gerla M (1998) GloMoSim: a library for parallel simulation of large-scale wireless networks. Simul Dig 28:154–161

Chapter 15
Topology Control

Abstract Topology control in ad hoc wireless networks aims to provide a sparser graph by preventing the use some of the existing communication links so that some of the neighbors of a node are excluded from its neighbor list deliberately, which would result in a simpler network graph. A hierarchical network structure is an effective way to organize a network comprising a large number of nodes. An efficient method of providing hierarchy and therefore scalability in ad hoc wireless networks is to group nodes of the network into *clusters*. Building such a hierarchy has many advantages including routing and load balancing. An efficient way of constructing a backbone and clusters is graph domination. In this chapter, we examine methods to construct sparse graphs, called *local graphs*, clustering, and the use of connected dominating sets for topology control.

15.1 Introduction

A sensor network may be densely deployed to provide redundancy for fault tolerance. Similarly, a mobile network may dynamically have a group of nodes located densely in a region. In such ad hoc networks, a node has many neighbors resulting in unwanted interference with them. Also, multi-hop and less energy consuming paths may be preferred instead of a direct and a more power consuming route to a destination. Under such conditions, a sparser and connected subgraph of the network graph may be used for communication. *Topology control* in wireless ad hoc networks restricts the allowed communication paths by obtaining a sparser subgraph of the network with less communication links and sometimes with less communicating nodes. Formally, given the network graph $G(V, E)$, a topology control method obtains $G'(V', E')$ such that $V' \subset V$ and $E' \subset E$.

Topology control may be achieved in various ways. One possible approach is to have only a subset of nodes active at any given time to conserve energy. In this case, a chosen subset of the neighbors of a node are considered as connected neighbors, and the rest are discarded. In another class of methods, only a subset of links are used for communication. The communicating neighbors may be chosen by adjusting the transmission power of a node.

Backbone construction is an effective method for topology control that provides a significant reduction in the number of messages communicated in an ad hoc wire-

K. Erciyes, *Distributed Graph Algorithms for Computer Networks*,
Computer Communications and Networks, DOI 10.1007/978-1-4471-5173-9_15,
© Springer-Verlag London 2013

Fig. 15.1 Backbone example

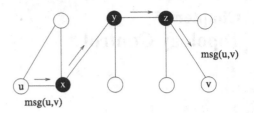

less network. When a backbone is constructed, a message from a source is first
routed to the nearest neighbor backbone node, then along the backbone to the backbone node closest to the destination, and finally to the destination as shown in
Fig. 15.1. The sender of the message is node u that simply sends the message to
node x, which is its backbone neighbor. Any node on the backbone examines the
destination in the message, and if this destination is not one of its neighbors, it floods
the message to other backbone nodes other than the sender. Finally, node z finds that
the destination node v in the message header is its neighbor, so it passes the message
to v. In this so-called *backbone-based routing*, the number of update messages for
routing is also significantly reduced. A clear requirement is that the backbone nodes
should be connected and that every node should be in the backbone or a neighbor to
a backbone node.

Backbones in ad hoc wireless networks can be constructed by clustering algorithms that partition the network into a number of clusters, each with a clusterhead
(CH), and the backbone then consists of these CHs and nodes that connect them.
Alternatively, a connected dominating set (CDS) of the network can be constructed,
where every node is at most one hop away from a node in the CDS. This way, each
node in the CDS is a member of the backbone. Constructing a CDS also allows an
indirect method for clustering such that each node in CDS is a CH of the nodes it
dominates. We will first review locally defined graphs and then investigate direct
clustering algorithms and construction of backbones using connected dominating
sets for WSNs and MANETs in this chapter.

15.2 Desirable Properties

Assuming that the topology control algorithm provides a graph $G'(V, E') \in$
$G(V, E)$, the following are desirable in G'.

15.2.1 Connectivity

A basic and important requirement from any topology control algorithm is that G'
should be connected, that is, there should be a path between any vertex pair u, v
in G'. Some algorithms provide *k-vertex-connectivity* or *k-edge-connectivity*, which
means that at least k vertices or edges must be removed from G' to disconnect it.

15.2.2 Low Stretch Factors

For a weighted graph $G(V, E, w)$, the distance between two vertices u and v is defined as the sum of the weights of the shortest path between them. The graph $G'(V, E')$ is called a k-*spanner* of G if every vertex pair u and v of G' is at most k times more distant than their distances in G. Formally, G' is a k-spanner of G if for all $u, v \in V$, $d'_G(u, v) \leq k d_G(u, v)$. A *stretch factor* is the largest ratio of a quantity measured in a subgraph to the same quantity measured in the original graph. If weights are assigned to reflect the distances of links, the stretch factor is called the *distance stretch factor*. Formally,

$$distance\ stretch\ factor = \max_{u,v \in V} \frac{d'_G(u, v)}{d_G(u, v)}, \tag{15.1}$$

where $d'_G(u, v)$ and $d_G(u, v)$ are the distances between the nodes u and v in G' and G, respectively. If the weight assigned to a link $\{u, v\}$ represents the minimum power required to communicate over this link, then the stretch factor is called the *power stretch factor*, which is specified as follows:

$$power\ stretch\ factor = \max_{u,v \in V} \frac{E'_G(u, v)}{E_G(u, v)}, \tag{15.2}$$

where $E'_G(u, v)$ and $E_G(u, v)$ are the minimum powers required for communication between the nodes u and v in G' and G, respectively. The *hop stretch factor* is the largest ratio of the minimum numbers of hops in G' and G between any two vertices in V as follows:

$$hop\ stretch\ factor = \max_{u,v \in V} \frac{H'_G(u, v)}{H_G(u, v)}, \tag{15.3}$$

where $H'_G(u, v)$ and $H_G(u, v)$ are the hop distances between the nodes u and v in G' and G, respectively. Having large stretch factors means that communication between nodes in G' has become more difficult than in G. For this reason, a general desirable property from any topology control algorithm is that the stretch factors should be as low as possible.

15.2.3 Bounded Node Degree

Most of the topology control algorithms attempt to provide a sparser graph. However, the sparser graph G' may still have high-degree vertices. A high-degree vertex in G' means that it has many neighbors resulting in dense communication around it and also interference problems with its neighbors. Another requirement from a topology control algorithm therefore is that the degree of the vertices should be bounded by a constant k such that there are no vertices of degrees higher than k.

Low interference between neighbors is also required to prevent collisions between the communicating neighbors. A planar graph has its edges intersecting only in the vertices, and planarity is therefore another requirement from a topology algorithm as it results in lower interference than an arbitrary graph with intersecting edges. We will look into ways of achieving sparser networks by eliminating some of the links from the network by disabling communication to some of the neighbors of the nodes in the next sections.

15.3 Locally Defined Graphs

In most of the distributed algorithms for topology control, nodes use their one- or two-hop neighbors connectivity to construct sparser graphs. In the following sections, we describe the main geometric structures for topology control and algorithms to obtain these structures based on the local neighbor information. Three procedures that are common to most of the algorithms that will be investigated are shown in Algorithm 15.1. The first procedure *find_neighs*1 sends a probe message to all neighbors of node *u* to find their distances assuming that the MAC layer does not provide this information. The *probe* message is time stamped with t_1, and when all the replies are received, the time stamps t_2 from the receivers can be used to find the physical distances assuming that each node responds to a *probe* message by a time-stamped *reply* message and nodes are time synchronized. This procedure may be activated periodically or when needed to estimate the distances of the nodes. A set of tuples in the form $\langle node\_id, node\_distance \rangle$ is kept in *dist* data structure, and any responding neighbor to the *probe* message is included in this structure. The second procedure *find_neighs*2 finds neighbors of neighbors by forming lists of neighbors (*all_dists*), which are sets of tuples $\langle j, dists \rangle$, where *dists* is the list of all neighbors of node *j*. It is assumed that each node *j* receiving a *request* message responds by the *info*(*j*, *dists*) message that includes its list of neighbors.

15.3.1 Nearest-Neighbor Graphs

In the *Nearest-Neighbor Graph* (*NNG*), each vertex is connected to its nearest vertex. Assuming that ties are broken by unique identifiers, vertex *u* has a single closest vertex to it, however, *u* can be the closest vertex to another vertex. Since being nearest is not symmetric, the NNG produces a directed graph as shown in Fig. 15.2. NNG in general is not connected due to its sparsity, and for this reason, it is not used for topology control. It produces planar graphs that are not spanners.

Each vertex *u* is connected to its *k* nearest neighbors in the *k-Nearest-Neighbor Graph* (*k-NNG*). A topology that is better connected than *NNG* can be obtained if only bidirectional edges that have end vertices as the *k*-closest neighbors of each other are included in *k*-NNG. For randomly nodes in a square and a constant λ, the resulting graph is connected with high probability if $k \leq \lambda(\log n)$.

Algorithm 15.1 Procedures for finding and allocating neighbors

```
 1: set of tuples dists, all_dists ← ∅
 2: int dist
 3: message types probe, reply, info, request
 4:
 5: procedure find_neighs1(dists)
 6:     broadcast probe(i, t1)                                    ▷ probe neighbors
 7:     while ¬timeout do
 8:         receive reply(j, t2)
 9:         dist_j ← vel/(t2 − t1)                                ▷ calculate distance of j
10:         dists ← dists ∪ (j, dist_j)                          ▷ include j and its distance in dists
11:     end while
12: end procedure
13:
14: procedure find_neighs2(all_dists)
15:     broadcast request                                        ▷ probe neighbors
16:     while ¬timeout do
17:         receive info(j, dists_j)
18:         all_dists ← all_dists ∪ (j, dists_j)
19:     end while
20: end procedure
21:
```

Fig. 15.2 Nearest neighbor
graph example

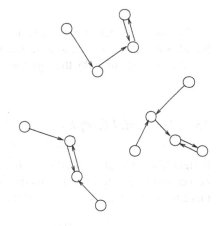

A possible and unsymmetric implementation of the k_NNG algorithm that finds k_NNG neighbors is shown in Algorithm 15.2, which first calls procedure *find_neighs1* to find the neighbors of node u from which the closest k nodes are determined. The algorithm then proceeds by broadcasting the *notify* message, which informs the chosen k nodes that they are neighbors. This second step is necessary as node $v \in \Gamma(u)$ may not have u as the closest neighbor as closeness is not symmetric, and hence will not be aware that it is a neighbor to u unless notified.

Algorithm 15.2 *k_NNG_Const*

1: **set of int** *all_neighs*, *k_neighs*
2: *find_neighs*1(*all_neighs*)
3: **sort** *all_neighs* in ascending order
4: *k_neighs* ← the first *k* neighbor nodes of *all_neighs*
5: **broadcast** *notify*(*k_neighs*)

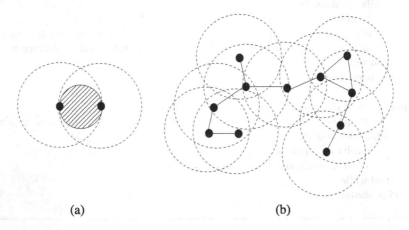

(a)   (b)

Fig. 15.3 Gabriel graphs: (**a**) the empty circle between two nodes; (**b**) an example network

The total number of messages in *k_NNG_Const* algorithm is 2*n* as each node broadcasts exactly two messages: first, to find neighbors by the *probe* message and, second, to notify them by the *myneighs* message.

15.3.2 Gabriel Graphs

In the Gabriel Graph (*GG*), there exists an edge between vertices *u* and *v* if there are no vertices in the disk that has its diameter as the edge between *u* and *v*. Formally, there is an edge between *u* and *v* if there is no $w \in V$ such that

$$d(u, w)^2 + d(v, w)^2 \le d(u, v)^2. \tag{15.4}$$

Since this relation is symmetric, *GG* provides a bidirectional and a connected graph as shown in Fig. 15.3.

The algorithm *GG_Const* checks this condition to include a node in the GG as shown in Algorithm 15.3. It first finds the neighbors of vertex *u* by calling procedure *find_neighs*1 to find its immediate neighbors and then calls *find_neighs*2 to find the two-hop neighbors. Using this information, the intersection set between vertices *u* and *v* can be found. The algorithm then tests the distance equation for each element

---

Algorithm 15.3 *GG_Const*

```
 1: set of int neighs ← ∅
 2: set of tuples dists, all_dists ← ∅
 3: find_neighs1(dists)                              ▷ determine one-hop neighbors
 4: find_neighs2(all_dists)                          ▷ determine two-hop neighbors
 5: find distances between each pair of nodes
 6: for all v ∈ Γ(u) do                                    ▷ Check GG condition
 7:     for all w ∈ (Γ(u) ∩ Γ(v)) do
 8:         if d(u, w)² + d(v, w)² ≤ d(u, v)² then
 9:             neighs ← neighs ∪ {v}       ▷ if satisfied, include node in neighbors
10:         end if
11:     end for
12: end for
```
---

in each intersection set. If this equation is satisfied, the neighbor v is included in the neighbor set, and any neighbor that satisfies this test is included in the neighbor list.

Total number of messages in *GG_Const* algorithm is $2n$ as each node broadcasts exactly two messages: first, to find one-hop neighbors by the *probe* message and, second, to find two-hop neighbors by the *request* message.

15.3.3 Relative Neighborhood Graphs

In the Relative Neighborhood Graph (*RNG*), two vertices u and v are connected if there are no other vertices that are closer to both u and v than the distance between u and v. In other words, there is no node w that exists in the intersection of the two disks centered at u and v as shown in Fig. 15.4. Formally, en edge $\{u, v\}$ in *RNG* implies

$$d(u, v) \leq max(d(u, w), d(v, w)) \quad \forall w \in (\Gamma(u) \cap \Gamma(v)). \quad (15.5)$$

RNG condition is symmetric, therefore the resulting graph is bidirectional and connected as shown in Fig. 15.4. A possible algorithm to form an RNG would work similar to Algorithm 15.3 with the only difference being the test condition to be included as a neighbor.

15.3.4 Delaunay Triangulation

A *Voronoi Diagram* for a set of points in the plane provides a number of cells where the borders of the cells are equidistant between the points in the adjacent cells. Each Voronoi Cell containing the hosting point p_i contains all the points in the plane that are closer to p_i than any other point in the plane. An example of Voronoi diagram

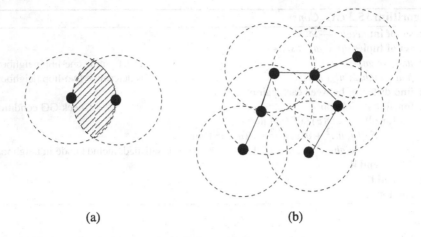

(a)   (b)

Fig. 15.4 Relative neighborhood graph example

Fig. 15.5 A Voronoi diagram
for nine points

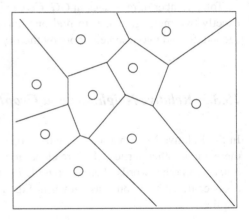

is shown in Fig. 15.5. Formally, given a set $P = \{p_1, p_2, \ldots, p_n\}$ of n points in a plane, a point q is included in the cell corresponding to $p_i \in P$ if and only if

$$d(q, p_i) < d(q, p_j) \quad \forall p_i \in P, j \neq i. \tag{15.6}$$

The Voronoi diagrams have many applications in astronomy, computer science, and bioinformatics. In *Delaunay Triangulation (DT)*, two vertices are connected if there is a circle passing through these vertices. DT provides a dual graph of a Voronoi diagram such that a DT graph has a vertex for every Voronoi cell and has an edge between the two vertices if the corresponding cells share an edge. A Gabriel Graph has an edge $\{u, v\}$ between two vertices u and v if there is an any empty disk between these two vertices. Based on this condition, GG ⊆ DT as GG is a special case of DT. Assuming that there may be maximum three points on a circle, all faces of a DT graph are triangles as shown in Fig. 15.6, which is the DT of Fig. 15.6.

Fig. 15.6 A Delaunay Triangulation of the graph in Fig. 15.5

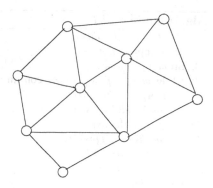

Fig. 15.7 A Yao Graph example with $k = 8$

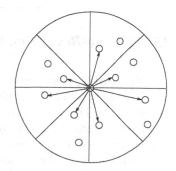

15.3.5 Yao Graphs

The *Yao Graph* (YG_k) divides the plane into k cone-shaped regions with $k \geq 6$ originating from the source node u. The closest node v, if any, to node u is chosen from each region and a directed link from u to v is formed as shown in Fig. 15.7. In case of a tie, the node with the lowest identifier is chosen. As with the NNG, the Yao Graph YG_k is directed as being closest in a cone of a node is not symmetric. A similar structure, called the *O Graph*, uses the shortest projection to the x-axis instead of the Euclidian distance [24]. The YG_k has a distance stretch factor $1/(1 - 2\sin(\pi/k))$ and a power stretch factor $1/(1 - 2\sin(\pi/k)^\beta)$ [19]. The maximum vertex degree Δ of YG_k is $n - 1$.

15.3.5.1 Symmetric Yao Graph

Li et al. [20] proposed Symmetric Yao Graphs YGS_k where an edge is included in YGS_k if and only if both edges $\{u, v\}$ and $\{v, u\}$ exist in the Yao Graph YG_k. In this case, the maximum node degree is k [6]. They also showed that the graph $YGS_k(V)$ is strongly connected if $UDG(V)$ is connected and $k \geq 6$. The experiment by Li et al. also showed that $YGS_k(V)$ has a small power stretch factor in practice.

Algorithm 15.4 *CBTC*

1: **set of int** *neighs*, $dirs_u \leftarrow \varnothing$
2: **set of tuples** *dists* $\leftarrow \varnothing$
3: **int** $pow_u \leftarrow pow_{\min}$
4: **while** $pow_u \leq pow_{\max} \wedge \exists \alpha$-cone in $dirs_u$ **do**
5:     $pow_u \leftarrow pow_u + step$
6:     *find1_neighs(dists)*
7:     $v \leftarrow min\{dists\}$
8:     *neighs* $\leftarrow$ *neighs* $\cup \{v\}$
9:     *dirs* $\leftarrow$ *dirs* $\cup \{dir_v\}$
10:    *dists* $\leftarrow \varnothing$
11: **end while**
12: **broadcast** *notify(neighs)*

15.3.6 Cone-Based Topology Control

In cone-based topology control, each node searches its neighbors in the region defined by a cone with a base angle α, and it forms a link with the first neighbor it finds in this region. This process continues until the area around a node is fully covered. Node i that starts the search by sending broadcast *probe* messages in a cone area defined by angle α. The transmission power is increased until the first neighbor j acknowledges the probe message and v is marked as a neighbor.

After a neighbor is marked in each cone, some of these neighbors are eliminated as follows. If some neighbor v discovered initially can be reached less costly via another neighbor p, v is discarded from the neighbor list. It was shown in [29] that for $\alpha = 2\pi/3$, the graph obtained is connected. The α value of $5\pi/6$ was later shown to be sufficient for connectivity [19]. Algorithm 15.4 describes the operation of the basic step of the CBTC where node i incrementally increases its power and waits for acknowledgements from neighbors for a period of time. Whenever an acknowledgement is received from node j, its distance $dist_j$ based on signal strength or timestamp value is calculated and stored in the *dists* structure as a tuple $\langle j, dist_j \rangle$. When node i reaches the timeout, it finds the minimum distance node v in the *dists* structure and assigns this node as its neighbor. This process is repeated until all directions around node i are covered [21].

15.4 Clustering

Given a graph $G(V, E)$, clustering divides the vertex set V into a collection of subsets $\{V_1, V_2, \ldots, V_k\}$, where $V = \bigcup_{i=1}^{k} V_i$, and each $V_i \in V$ induces a connected subgraph $G_i \in G$ that may overlap. An elected node in each cluster as the CH or the *leader* of the cluster is responsible for cluster management functions and providing hierarchical routing. CHs may be rotated among the members of the cluster to provide load balancing and fault tolerance. A *gateway* node is a node that connects two clusters. The three types of nodes in a clustered network therefore are the CHs, gateway nodes, and the ordinary nodes.

Clustering is NP-hard in general, and for this reason, various heuristics may be used to form clusters. Clustering algorithms determine the CH using some static property of the graph such as the identities or degrees of the nodes or a combination of these parameters to assign these special nodes and build a cluster around them. The main advantages to be gained by clustering an ad hoc wireless network are as follows:

1. A virtual backbone consisting of CHs and the gateway nodes provides an efficient method of routing in ad hoc networks as it forms a simple topology to maintain. This method of hierarchical routing results in smaller routing tables, which are stored by the CHs and the gateway nodes only, and hence efficiency of routing is improved.
2. Messages needed to update routing tables are reduced.
3. Usage of clustering also provides efficient MAC protocols by improving the throughput, scalability, and power consumption.

The two phases of clustering are the cluster formation and the cluster maintenance. Once clusters are formed, the cluster structure should be preserved as long as possible. In a MANET, reelection of a CH may often be necessary as the nodes frequently change their positions.

15.4.1 Clustering in Sensor Networks

Sensor networks consist of a large number of stationary nodes, usually deployed in hostile environments. Since changing batteries is in general very difficult in these environments, clustering algorithms in sensor networks should consider energy efficiency as one of the most important criteria.

A method to provide energy efficiency is to minimize total distances to the CH. The energy e to transfer a message between two nodes that are d distance apart is given by [15]

$$kd^c \quad (2 < c < 4), \tag{15.7}$$

where k and c are constants dependent on the wireless system. Ghiasi et al. [15] proposed an algorithm to minimize the sum of the squares of distances from all nodes in the cluster to CH. Another method is first clustering the nodes based on their distances, and then the assignment of the lowest power levels needed for intra-cluster communication to ordinary nodes and CHs and lowest power for inter-cluster communication to gateway nodes as in [25]. Kawadia et al. proposed to have clusters with different level of transmission powers [17], where each node may belong to more than one cluster of different power levels, and routing may be determined using the least power paths.

CHs consume more energy than ordinary nodes due to their relay position in the network. The power consumption of a CH may be improved by rotating the clusterhead function among the nodes of the cluster and providing balanced clusters such that the number of nodes in each cluster is similar. Liu and Lin [23] provided a scheme to elect the node with the highest residual energy in the cluster as CH periodically.

15.4.2 Clustering in MANETs

The cluster maintenance process is more important in a MANET as topology is dynamic. Selection of optimum number of CHs is NP-hard, and for this reason, node identifier, node degree, node mobility, and its battery power are frequently used for the selection of CHs heuristically, as noted before. In a MANET, a CH or an ordinary node of its cluster may move to a position so that they are out of range from each other. In such a case, a new CH should be searched by the ordinary nodes. Low-maintenance algorithms are usually preferred in MANETs as maintenance is frequently needed due to the random and frequent movement of the nodes.

A clique in graph $G(V, E)$ consists of a subset V' of V, where for all $v \in V'$, there exists an edge between v and all other vertices in V'. Krishna et al. [18] proposed a clustering method for a MANET where maximal cliques are used as clusters. There are no CHs in this algorithm, and gateway nodes are used to transfer messages similarly to Internet Border Gateway Protocol (BGP) routing [28].

15.4.3 Performance Metrics

We need to define few parameters to understand the goodness of the clustering.

Definition 15.1 (Cover) A *cover* of a graph $G(V, E)$ is a collection of clusters $C_1, C_2, \ldots, C_k$ such that $\bigcup_{i=1}^{k} C_i = V$.

Definition 15.2 (Partition) A *partition* of a graph $G(V, E)$ is a collection of disjoint clusters $C_1, C_2, \ldots, C_k$ that form a cover of G.

Definition 15.3 (Size of a cluster) The size $S(C)$ of a cluster C is the number of nodes in C.

Definition 15.4 (k-Cluster) A k-cluster is defined by a subset of nodes that are mutually reachable by a path of length at most k for some fixed k. A k-cluster with $k = 1$ is a clique.

Definition 15.5 (k-Hop cluster) A k-hop cluster is a set of nodes within at most k-hop distance from their CH.

Fig. 15.8 (a) A 3-cluster without a CH; (b) A 2-hop cluster with a CH

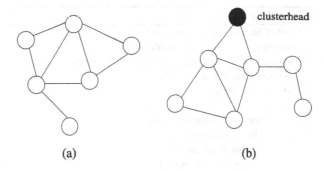

(a)   (b)

The difference between the last two definitions is whether the distance is considered between any pair of members in a cluster, or the CH and each member. By adjusting the parameter k, the number of clusters and CHs could be controlled. Figure 15.8 displays these concepts.

The quality of any clustering can be described by the clustering coefficients defined as follows:

Definition 15.6 (Node clustering coefficient) Given a simple, connected, and undirected graph $G(V, E)$ and a node $v \in V$ that has a neighbor set $N(v)$, assume that $n_v = |N(v)|$ and m_v is the number of edges in the subgraph induced by $N(v)$ in G. The *clustering coefficient $cc(v)$* for a node is the ratio of m_v to the maximum edges that v can have, that is,

$$cc(v) = \frac{2m_v}{n_v(n_v - 1)}. \tag{15.8}$$

Definition 15.7 (Graph clustering coefficient) The graph clustering coefficient $CC(G)$ of the graph $G(V, E)$ is the average value of the node clustering coefficients as follows:

$$CC(G) = \frac{1}{|V|} \sum_{v \in V} cc(v). \tag{15.9}$$

Two important aspects of a cluster are its locality and sparsity levels [27]. Locality level of a cluster can be determined by its radius and diameter, whereas average degree provides information about its sparsity.

15.4.4 Lowest-ID Algorithm

In the lowest identifier algorithm (*Low1_Clust*) due to Gerla and Tsai [14] to cluster nodes of a wireless ad hoc network, each node periodically broadcasts the nodes it can hear (detect) including itself in its UDG, after which the following rules are applied:

Algorithm 15.5 *Low*1_*Clust*

```
 1: set of int neighs, my_cheads ← ∅
 2: states chead, gateway, ordinary
 3: message types update, i_am_chead, ordinary
 4: boolean has_chead ← false
 5: state ← ordinary
 6: loop                                                    ▷ Do periodically
 7:     broadcast update(my_id)                          ▷ check active neighbors
 8:     while ¬timeout do
 9:         receive update(j)
10:         neighs ← neighs ∪ {j}
11:     end while
12:     if my_id = min{neighs} then
13:         broadcast i_am_chead
14:         state ← chead
15:     else broadcast ordinary                    ▷ needed to check all neighbors
16:     end if
17:     while received ≠ neighs do                      ▷ receive neighbor info
18:         receive msg(j)
19:         if msg(j).type = i_am_chead  then
20:             my_cheads ← my_cheads ∪ {j}
21:             if has_chead = false then has_chead ← true    ▷ CH marked first time
22:             else state ← gateway                       ▷ node is a gateway
23:             end if
24:         end if
25:         received ← received ∪ {j}
26:     end while
27: end loop
```

---

1. A node decides to be a CH if it does not hear a node with a higher identifier than itself.
2. The lowest identifier neighbor that a node hears is marked by the node as its CH, unless that node voluntarily gives up its position as a CH.
3. A node that hears two or more CHs becomes a *gateway* that joins two clusters.

Algorithm 15.5 shows a possible implementation of *Low*1_*Clust*. Each node first discovers its neighbors by broadcasting an *update* message. A protocol at MAC layer could already provide active neighbor identifiers, in which case lines 7–11 of the algorithm should be omitted; here we assumed a periodically invoked algorithm that checks active neighbors. When a node finds that it has the lowest identity among all neighbors, it broadcasts that it is the CH. Any node that finds that it has two broadcasting neighbors becomes a gateway. Every node is classified as an ordinary, CH, or a gateway node at the end of the algorithm.

Figure 15.9(a) shows an example network that is divided into three clusters with nodes as $C_1 = \{1, 5, 8\}$, $C_2 = \{2, 3, 7, 9\}$, and $C_3 = \{4, 6, 8, 9\}$ with CH nodes 1, 2, and 4, respectively. Nodes 8 and 9 are the gateway nodes between clusters C_1 and C_3 and between C_2 and C_3, respectively. The total number of messages communicated

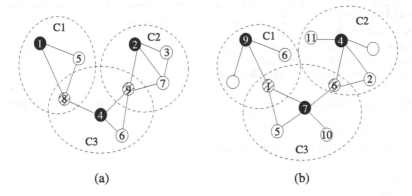

Fig. 15.9 (a) Lowest1, (b) highest degree algorithms clusterheads are shown in *bold*; gateway nodes are shown in *stripes*

in Algorithm 15.5 is $2n$ as each node will broadcast one message (*update*) to its neighbors and another message (*i_am_chead* or *ordinary*) to inform whether it is a CH or an ordinary node. The *Low1_Clust* algorithm may result in unstable CH selection. A low-identifier node entering an already formed cluster may have all of the structure of the cluster to be reorganized, whereas it could simply have joined the cluster.

15.4.5 Highest Connectivity Algorithm

Gerla and Tsai [14] further provided an algorithm we will call *High_Clust*, where the highest degree node among all neighbors becomes a CH, the neighbors of a such node are then covered, and the algorithm continues until all nodes are covered. Node u periodically broadcasts its identifier and the size of its open neighbor set $N(u)$ and becomes a CH if it is the most highly connected node among its neighbors. The degrees of nodes change rapidly in a MANET, causing frequent changes of CH in this algorithm. The following rules govern the operation of *High_Clust* algorithm:

- A node that has not elected its CH is an *uncovered* node, otherwise a node with a CH is a *covered* node.
- A node is elected as a CH if it has the highest degree and, therefore, the highest connectivity, among all of its uncovered neighbors. In the case of a tie, the lowest identity node becomes the CH.
- A node gives up its role as a CH if it elects another node as its CH.

   Figure 15.9(b) shows an example network that is divided into three clusters as $C_1 = \{1, 3, 6, 9\}$, $C_2 = \{2, 4, 6, 8, 11\}$, and $C_3 = \{1, 5, 6, 7, 10\}$ with CH nodes 9, 4, and 7, respectively.

   In both the lowest identity and the highest connectivity algorithms, CHs may not be directly connected to each other, and each CH is directly connected to every other

---

Algorithm 15.6 *Low2_Clust*

```
 1: set of int neighs ← ∅
 2: int i, j, state, my_clusid, clust_ids[dᵢ]
 3: states chead, gateway, ordinary
 4: message types update, found_clust, ordinary
 5: state ← ordinary
 6: if i = min{neighs} then                          ▷ if i have min id, declare CH
 7:     state ← chead
 8:     my_clusid ← i
 9:     broadcast found_clust(i, my_clusid)
10:     neighs ← neighs \ {i}
11: end if
12: repeat
13:     receive found_clust(j, cid)
14:     clust_ids[j] ← cid
15:     if (j = cid) ∧ ((my_clusid =⊥) ∨ (my_clusid > cid)) then   ▷ if new id is smaller
16:         my_clusid ← cid                           ▷ then set new CH
17:     end if
18:     neighs ← neighs \ {j}
19:     if i = min{neighs} then
20:         if my_id =⊥ then
21:             my_clusid ← i
22:         end if
23:         broadcast found_clust(i, my_clusid)       ▷ broadcast my CH
24:         neighs ← neighs \ {i}
25:     end if
26: until neighs = ∅
```

---

node in its cluster, forming 1-clusters. The intended role of CH is to perform MAC layer functions such as channel access, power measurements, and maintain time division frame synchronization. Any existing routing algorithm that is independent from the clustering can be used with these algorithms. Major drawbacks with this algorithm is that there may be numerous ties and it may not provide CHs in special graphs such as triangular graphs.

15.4.6 Lowest-Id Algorithm: Second Version

Lin and Gerla described a modified version of the lowest-ID algorithm (*Low2_Clust*) to solve the problems of the above algorithms [22] and partition a network into nonoverlapping clusters. They assumed that every node knows identifiers of their neighbor nodes, which may be provided by the physical layer. They further assumed that the network topology does not change during algorithm execution and a message is correctly received by all of the neighbors of a node in a finite time.

The algorithm starts by a node sending the *cluster*(*id*, *cid*) message to its neighbors that it has decided to create a cluster if it finds that it has the lowest identity

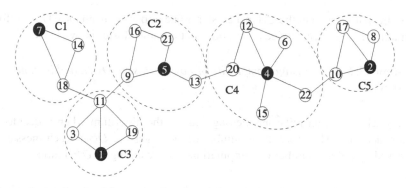

Fig. 15.10 *Lowest_ID2* algorithm example

among its neighbors, as shown in Algorithm 15.6. After this step, each node listens to their neighbors, and if a node hears a *cluster(id, cid)* message when its cluster identifier is not determined, it sets its cluster identifier to the incoming *cid* parameter, which is the identifier of the node that sent the *cluster* message. It also does so if *cid* is smaller than its current cluster identifier. If none of the lowest identifier neighbor nodes that have sent the *broadcast* message have not declared themselves as CHs, a node decides to create a cluster itself by broadcasting a *cluster* message. If one or more neighbors of a node have declared to be CHs, it chooses the lowest identifier of these nodes as its CH and broadcasts this decision.

In summary, after the initial lowest identifier messages, each node chooses a CH or decides to become a CH and broadcasts this decision. Each node sends exactly one message during the algorithm resulting in *n* messages in total.

For dynamic networks where nodes may arbitrarily join or leave the network, the following are also provided in *Low2_Clust*. Any incoming node to the network that can communicate with all nodes in a cluster in two hops is allowed to join that cluster. In the case of a lost link, the highest degree node becomes the new CH, and its neighbors are included as the cluster members. The former cluster members are allowed to join the new cluster or form a cluster of their own, which may result in single-node clusters necessitating additional procedures for merging or rearranging clusters [26]. Figure 15.10 shows an example network that is divided into five disjoint clusters by this algorithm, where the lowest identifier in each cluster is the CH.

15.4.6.1 Analysis

Since every node can determine its cluster, the *neighs* set will eventually be empty. and the algorithm will terminate.

Theorem 15.1 *Nodes in a cluster are at most 2-hops away from each other.*

Proof Each node except the CH in a cluster is one hop away from the CH; therefore, the maximum distance between two nodes in the same cluster is two hops.   □

Theorem 15.2 *The message complexity of Low2_Clust is $O(n)$, and its time complexity is also $O(n)$.*

Proof Each node transmits one message during the algorithm when it decides its cluster identifier. Therefore, total number of messages are $O(n)$. Each message is processed by a fixed number of computation steps resulting in $O(n)$ time.   □

This algorithm may be used to guarantee the QoS requirements such as the bandwidth and delay for multimedia applications in MANETs. The network can be partitioned into clusters so that bandwidth sharing by the clusters can be achieved and virtual circuits with QoS guarantee can be established [22].

Basagni et al. [6] proposed to use weights of nodes instead of the lowest identifier of *Low2_Clust* to elect the CHs where the weight of a node depends on its mobility. Basagni also proposed a generalization of the algorithm in [6] by allowing each CH to have maximum of k neighboring CHs [7]. Chatterjee et al. [6] also used weights for nodes that consist of node's degree, sum of distances to all neighbors, speed of node, and the cumulative time node serves as a CH.

15.4.7 k-Hop Clustering

The algorithm to find k-hop clusters due to Nocetti et al. [26] (*Conid_Clust*) generalizes *Low2_Clust* algorithm to find k-clusters by implementing this algorithm for k-hops. It is assumed that all nodes are aware of their k-hop neighbors. Substituting 1-hop with k-hops would yield an algorithm as follows. A node that finds that it has the lowest identity among all its k-hop neighbors broadcasts its intent to cerate a cluster to all of its k-hop neighbors. As in the *Low2_Clust* algorithm, a node chooses the lowest identity k-hop neighbor as its CH and broadcasts this decision to k-hop neighbors. If none of the k-hop neighbors of a node has declared itself as a CH, it can start to create a cluster of its own.

Each node sends exactly one message about its clustering decision. The *Conid_Clust* algorithm considers both the node identity and degree when electing a CH as the algorithm may produce more clusters than needed. This algorithm may produce wrong results because of ties. Each node is identified by the pair $\langle d, ID \rangle$, where d is the connectivity of the node. The *clusterhead priority* of a node is determined by these parameters, and instead of the lowest identity in the above algorithm, *Conid_Clust* uses the highest clusterhead priority to elect clusterheads. The maintenance procedures of *Conid_Clust* are also modified to allow the maintenance of k-hop clusters.

Fig. 15.11 Operation of
ST_Clust in a sensor network

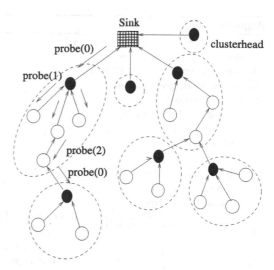

15.4.8 *Spanning-Tree-Based Clustering*

Erciyes et al. [12] provided an asynchronous single initiator spanning-tree-based clustering algorithm (*ST_Clust*) for WSNs, which is a modified version of the spanning tree construction algorithm (*Flood_ST*) of Sect. 4.2. They assumed that the sensor nodes are distributed randomly and the sensor field can be mapped to a two-dimensional space and all the sensor nodes have identical and fixed transmission ranges.

The algorithm proposed is described informally as follows. The sink node of the sensor network periodically starts the algorithm by sending a *probe*(0) message to its neighbors. Any node i that receives a parent message for the first time, sets sender as its parent, sends *ack* message to its parent and sends a *probe*(i, *n_hops*) message to all of its neighbors. The depth of subtree (*ds*) is provided as the modification to the above classical algorithm to form a spanning tree. Every node that is assigned a parent sets *n_hops* to (*n_hops* + 1) MOD *ds* and appends *n_hops* to its outgoing *probe* message. The recipient of the message with *n_hops* = 0 are the CHs, and *n_hops* $\leq d$ are *intermediate* or *leaf* nodes depending on their level within a subtree. The algorithm provides CHs as the subroots of the subtrees. Algorithm 15.7 shows the operation of the algorithm, and Fig. 15.11 displays the clusters constructed in a sensor network using this algorithm with *ds* = 2.

Theorem 15.3 *The time complexity of ST_Clust is $O(d)$, where d is the diameter of the network, and its message complexity is $O(n)$.*

Proof The time required for the algorithm is the largest distance between any two nodes which is the diameter d of the network. As the final spanning tree will have n nodes and $n - 1$ edges and each edge will have been traversed twice by *probe* and *ack* or *reject* messages, the total number of messages will be $O(n)$. $\square$

Algorithm 15.7 *ST_Clust*

1: **int** *parent* ←⊥; *my_chead*, *n_hops*
2: **set of int** *childs*, *others* ← ∅
3: **message types** *probe*, *ack*, *reject*
4: **states** *chead*, *intermed*, *leaf*
5: **if** *i* = *root* **then**   ▷ root initiates tree construction
6:     **send** *probe*(0) to *Γ(i)*
7:     *parent* ← *i*
8: **end if**
9:
10: **while** (*childs* ∪ *others*) ≠ (*Γ(i)*\{*parent*}) **do**
11:     **receive** *msg(j)*
12:     **case** *msg(j).type* **of**
13:             *probe(cid, n_hops)*:   **if** *parent* =⊥ **then**   ▷ *probe* received first time
14:                                     *parent* ← *j*
15:                                     **send** *ack* to *j*
16:                                     **if** *n_hops* = 0 **then**   ▷ *i* am the clusterhead
17:   *state* ← *chead*
18:   *cid* ← *i*
19:                                     **else if** *n_hops* = *ds* **then**
20:   *state* ← *leaf*
21:                                     **else** *state* ← *intermed*
22:                                     *my_chead* ← *cid*
23:                                     *n_hops* ← (*n_hops* + 1) MOD *ds*
24:                                     **send** *probe(cid, n_hops)* to *Γ(i)* \ {*j*}
25:                                     **else send** *reject* to *j*   ▷ *probe* received before
26:             *ack*:   *childs* ← *childs* ∪ {*j*}   ▷ include *j* in children
27:             *reject*:   *others* ← *others* ∪ {*j*}   ▷ include *j* in unrelated
28: **end while**

Banerjee and Khuller [5] also proposed a protocol based on a spanning tree by grouping branches of a spanning tree into clusters of an approximate target size.

15.5 Connected Dominating Sets

A convenient way of constructing a backbone is the building of a connected dominating set (CDS) so that for every node $u \in V$ of the graph $G(V, E)$, either $u \in$ CDS or is adjacent to a node in the CDS. In general, CDS algorithms for ad hoc networks can be maximal independent set (MIS) based or non-MIS based. Many algorithms to construct a CDS are based on MIS, and these algorithms further can be classified as *one-* or *two-phase algorithms*. In one-phase algorithms, the MIS nodes and the intermediate nodes are determined simultaneously. In two-phase algorithms, an MIS is first constructed in the first phase, and some optimal nodes are used to connect the MIS to get a CDS in the second phase. The selection of an intermediate node is usually based on the identifier of the node, its degree, or its residual energy. In this

---

Algorithm 15.8 *Seq_CDS*

1: Input $G(V, E)$
2: $S \leftarrow V, MIS \leftarrow \varnothing$
3: **while** $S \neq \varnothing$ **do**
4:     **pick** an arbitrary vertex $u \in S$
5:     $S \leftarrow S \setminus \{u \cup \Gamma(u)\}$
6:     $MIS \leftarrow MIS \cup \{u\}$
7: **end while**
8: $CDS \leftarrow MIS$
9: **for all** $u \in CDS$ **do**
10:     **for all** $v \in CDS \wedge$ two hop neighbor of u **do**
11:         **find** w between u and v with the highest degree
12:         $CDS \leftarrow CDS \cup \{w\}$
13:     **end for**
14: **end for**
15: **prune** any $u \in CDS$ that is redundant

---

section, we will review some representative algorithms for backbone construction, and we will assume that these algorithms operate in UDG environments.

15.5.1 A Sequential Algorithm using MIS

As the first step, we will describe a sequential algorithm that first finds an MIS and connects the nodes in this MIS to get a CDS in the second step. The algorithm, called *Seq_CDS*, obtains an MIS of $G(V, E)$ by arbitrarily choosing vertex $u \in V$, including u in the MIS and deleting u and all of its neighbors from G until there are no vertices of G left in the first phase as was described in Sect. 10.2; as shown in Algorithm 15.8. Connecting nodes of the MIS can be performed using various methods. For example, a node with the highest degree between any two nodes of the MIS can be used to connect them, and pruning can be performed as the final step to remove any redundant nodes from the CDS.

Figure 15.12 shows the operation of *Seq_CDS* in a sample graph of nine nodes with identifiers $1, \ldots, 9$. The first phase of the algorithm selects vertices 2, 4, and 9 arbitrarily to form the MIS = $\{2, 4, 9\}$ as shown by black nodes, and this phase concludes by including all the vertices of MIS in the CDS as shown in (a). In the second phase, intermediate vertices 1 and 6 with highest degrees between neighbor MIS nodes are used to connect these vertices to obtain the CDS in (b). The example pruning rule used here is the removal of a node from the CDS if it has a CDS neighbor with a higher identifier and this neighbor covers all of its neighbors. Node 4 is removed from the CDS in (c) as it satisfies this condition.

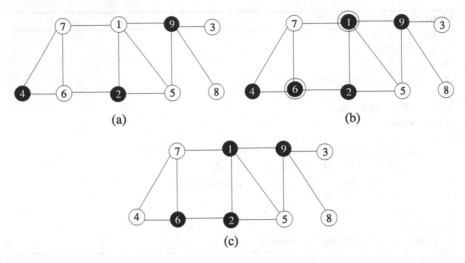

Fig. 15.12 CDS construction from MIS

15.5.2 Greedy Distributed Algorithms

Das et al. [10] provided two algorithms that are the distributed versions of Guha–Khuller algorithms we have seen in Chap. 11. In the first algorithm, nodes are assigned weights as their effective degrees, which are the numbers of their non-CDS neighbors. Initially, a small dominating set S is formed, which may have several disconnected components. This forest consisting of the edges $\{v_1, v_2\}$, where $v_1 \in S$ and $v_2 \in \Gamma(v_1)$, is then connected in the second stage using a distributed minimum spanning tree (MST) algorithm. The CDS obtained consists of the nonleaf nodes of the MST formed. This algorithm provides an approximation ratio of $2H\Delta + 1$ in $O(n + |C|\Delta)$ time using $O(n|C| + m + n \log n)$ messages where $|C|$ is the size of the CDS [10].

In the second algorithm, one- or two-step paths emanating from the current CDS are investigated to find the node with the greatest number of white nodes in each round. A node or a pair of nodes with the highest number of span is added to the existing CDS as in [16]. This algorithm achieves an approximation ratio of $2H\Delta$ in $O(|C|(\Delta + |C|))$ time using $O(n|C|)$ messages [10].

15.5.3 MIS-Based Distributed CDS Construction

Alzoubi et al. proposed a MIS-based CDS construction algorithm (*Alzoubi_CDS*) based on UDGs. Due to UDG imposed geographic constraints, a node can be adjacent to at most five independent neighbors. This algorithm consists of the following phases [1]:

1. Leader Election

2. Level Calculation
3. Color Marking

15.5.3.1 Leader Election

In this phase, a distributed leader election algorithm such as in [8] is used to construct a spanning tree T rooted at the leader. A root node may also be arbitrarily chosen, and this root node may construct a spanning tree using any of the algorithms we have seen in Part I. Convergecast of control messages can be used to notify the root that the first phase is over.

15.5.3.2 Level Calculation

When the first phase is over, the root starts the second phase by sending its level (0) to its children, which increase the level by one and send the new level to their children. Each node in T also records the levels of their neighbors. When the leaves compute their level, a convergecast operation is performed by the *complete* messages to the root over T, and the root knows that the second phase is over. Each node in the tree has a rank identified by the ordered pair of its level and its identity at the end of this phase, and the number of messages sent is $O(n)$.

15.5.3.3 Color Marking

In this phase of the algorithm, the aim is to have colored black all nodes that are the nodes in CDS or gray all nodes that are the neighbors of the nodes in CDS. The messages in the MIS construction phase are *dominator* sent by a MIS (black) node to its children and *dominatee* sent by a non-MIS (gray) node. Initially, all nodes are white and this phase of the algorithm is initiated by the lowest rank node which is the root, by broadcasting a *dominator* message. The ranked nodes now implement the following rules [1]:

1. A white node receiving a *dominator* message for the first time marks itself gray and broadcasts a *dominatee* to show that it has been dominated.
2. A white node that has received *dominatee* messages from all of the lower-rank neighbors enters MIS by marking itself black, sends a *dominator* message to all of its neighbors, and assigns its parent in T as its dominator.

The second phase finishes when the leaves of the tree are marked. In the final phase, the *invite* and *join* messages are used to connect the MIS formed in the previous phase to form a CDS. Initially, the root node sends the *invite* message to its MIS neighbors two hops away, and any black node that receives this message then joins the CDS together with the gray node that has sent it. This node then broadcasts the *invite* message to its neighbors. This algorithm has a time complexity of

$O(n)$, message complexity of $O(n \log n)$, and the resulting CDS has a size of at most $8OPT + 1$ [1]. Alzoubi et al. [2] also provided a multi-leader CDS algorithm to reduce communications and decrease the size of the CDS.

15.5.4 Pruning-Based Algorithm

The Wu and Li distributed algorithm (*Wu_MCDS*) finds an MCDS of a general graph in two phases. In the first phase, each node sends the identifiers of all its neighbor nodes to all its neighbors. At the end of the first phase, a node decides to be a potential member of the CDS by changing its color to black only if it has two nonadjacent neighbors. Each node that nominates itself to be black notifies its neighbors in the second phase. At the end of the second phase when all messages are received, two pruning rules are applied to reduce the size of the CDS by deleting redundant nodes from it. A node first checks whether a neighbor CDS node in its closed neighborhood with a higher identifier covers its entire neighbor set, in which case it exits the CDS. Secondly, if the open neighbor set of a node is covered by two CDS neighbors with higher identifiers, it backs off from CDS by changing its color to white. Formally, the rules for a node v are as follows:

- **If** $\exists u \in N[v] | (color_u = black) \wedge (N[v] \subseteq N[u]) \wedge (id_v < id_u)$ **then** $color_u \leftarrow white$
- **If** $\exists u, w \in N(v) | (color_v = color_u = color_w = black) \wedge (N(v) \subseteq (N(u) \cup N(w))) \wedge (id_v = min\{id_v, id_u, id_w\})$ **then** $color_v \leftarrow white$

Algorithm 15.9 shows a possible implementation of *Wu_MCDS*, where, unlike the original algorithm, a node broadcasts its color even if it is not in the CDS for proper synchronization. As each node sends exactly two messages in this implementation, the total number of messages transmitted is $2n$.

Figure 15.13 shows an example network where a CDS is formed at the end of first phase in (a), with nodes 3, 5, 6, and 4 as they all have two unconnected neighbors. Pruning with Rule 1 provides nodes 3 and 4 to change their colors to white as their closed neighborhoods (2, 3, 5, 6 and 1, 4, 5, 6) are covered by larger identifier CDS neighbor nodes as shown in (b).

Figure 15.14 displays pruning by Rule 2, where nodes 4, 1, 2, and 5 are marked in the first phase as they all have two unconnected neighbors. In the second phase, node 2 finds that its open neighborhood (nodes 3, 4, 1, 5, 6) are covered by the union of the open neighborhood of its neighbor 4 and 5, so it decides to change its color to white as shown in (b).

Cokuslu and Erciyes [9] extended the *Wu_MCDS* algorithm by involving degree of the nodes during pruning process. In their algorithm, only nodes that have isolated neighbors are marked black permanently in the first phase. Any other node that has two unconnected neighbors is colored gray, and the rest of the nodes do not change their colors.

Algorithm 15.9 *Wu_MCDS*

```
 1: int i, j, k; mycolor ← white
 2: set of int neighs1, neighs2 ← ∅
 3: message types phase1, phase2
 4: boolean terminated, phase1over
 5: send phase1(Γ(i)) to Γ(i) {Start Phase 1}
 6: while terminated ≠ true do
 7:     receive msg(j)
 8:         case msg(j).type of
 9:             phase1(Γ(j)):      neighs1 ← neighs1 ∪ j
10:                                2neighs[j] ← Γ(j)
11:             phase2(color):     neighs2 ← neighs2 ∪ j
12:                                neighcolors[j] ← color
13:
14:     if neighs1 = Γ(i) then                                    ▷ Phase 1 over
15:         if {∃j, k ∈ Γ(i)|j ∉ Γ(k)} then
16:             mycolor ← black
17:             broadcast phase2(mycolor)
18:             phase 1 over ← true
19:         end if
20:     end if
21:     if neighs2 = Γ(i) ∧ phase 1 over then                     ▷ Phase 2 over
22:         if mycolor = black then
23:             if (∃j ∈ Γ(i)|color_j = black) ∧ (Γ[j] ⊆ Γ[i]) ∧ (i < j) then   ▷ Rule 1
24:                 mycolor ← white
25:             end if
26:             if (∃j, k ∈ Γ(i)|color_j = color_k = black) ∧ (Γ(i) ⊆ (Γ(j) ∪ Γ(k))) ∧ id_v =
         min{id_v, id_u, id_w} then                              ▷ Rule 2
27:                 mycolor ← white
28:             end if
29:             terminated ← true
30:         end if
31:     end if
32: end while
```

In the second phase, they consider the degree of a node when marking it as black since a node with a higher degree should have a better chance of being in CDS. They implemented the following pruning rules during the second phase of the algorithm:

1. $\exists u \in N(v)$ which is marked *black* such that $N[v] \subseteq N[u]$;
2. $\exists u, w \in N(v)$ which is marked *black* such that $N(v) \subseteq N(u) \cup N(w)$;
3. $\exists u \in N(v)$ which is marked *gray* such that $N[v] \subseteq N[u]$ and $degree(v) < degree(u)$ or ($degree(v) = degree(u)$ and $id(v) < id(u)$);
4. $\exists u, w \in N(v)$ which is marked *gray* or *black* such that $N(v) \subseteq N(u) \cup N(w)$ and $degree(v) < min\{degree(u), degree(w)\}$ OR $degree(v) = min\{degree(u), degree(w)\}$ and $id(v) < min\{id(u), id(w)\}$;

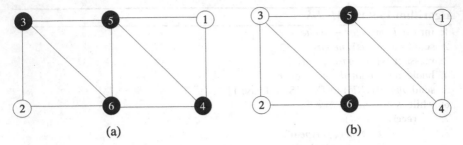

Fig. 15.13 *Wu_MCDS* pruning Rule 1 example

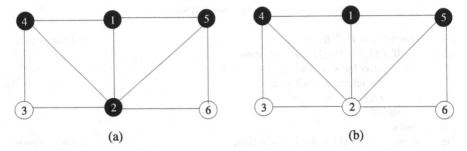

Fig. 15.14 *Wu_MCDS* pruning Rule 2 example

These rules favor nodes with higher degree to be included in the MCDS, and identifiers are used to break ties. Cokuslu et al. compared this algorithm to *Wu_MCDS* experimentally and showed that it provides MCDSs with significantly smaller sizes for a wide range of number of nodes.

15.6 Chapter Notes

Topology control using local graphs is a widely used method to provide sparser graphs with less communication links. This approach also prevents interference of simultaneous transmissions to some extent; however, we have seen that there are more efficient methods to reduce interference in Sect. 14.4.3. The main usage of local graphs in this context is for energy efficient communications rather than for prevention of interference. Locally defined graph-based topology control remains an active research topic in MANETs and WSNs.

We have seen few sample algorithms for clustering in ad hoc wireless networks. A survey of clustering in MANETS is given in [31], and a recent survey in [4]; a survey of clustering in ad hoc networks is presented in [30]. A clear distinction exists between a stationary, energy-sensitive sensor network, which consists of a large number of sensing nodes, and a MANET, where mobility of nodes is of primary concern. Clustering algorithms for sensor networks should aim at energy efficiency,

Fig. 15.15 Sample graph for
Exercise 1

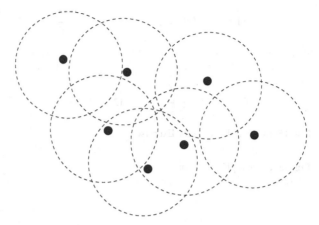

and such methods include minimizing the total distance to CH, rotating CHs, and
providing clusters of nodes having same power levels. In a MANET, power is in
general, not of primary, concern as nodes can be recharged and there is not a special
sink node as in a sensor network. Clustering in a MANET should aim at mobility,
and for this reason, low-maintenance algorithms such as the ones we have seen are
suitable for MANETs.

CDS-based backbone construction in ad hoc wireless is a thoroughly researched
area, and related publications are numerous. We have described few fundamental
distributed algorithms to provide a backbone using this method. Recently, there has
been a significant interest in finding d-dominating sets. Given a graph $G(V, E)$, a
d-dominating set S is a subset of vertices of G such that every $v \in V$ is at most
d hops away from at least one of the nodes in S. Finding minimum d-dominating
sets is an NP-hard problem [13]. The d-dominating sets have various applications
such as multicast systems, message routing, and placement of routers in computer
networks. Min–Max D-Clustering algorithm presented in [3] provides clusters by
selecting CHs based on both low and high identifiers, and it also results in d-hop
dominating sets.

15.6.1 Exercises

1. Construct the *NNG*, *GG*, and *RNG* networks for the nodes in the sample graph of
   Fig. 15.15, where each node has a transmission range shown by a circle around
   it.
2. Compare clustering in WSNs and MANETs in terms of objectives and implementations.
3. Show the clusters, CHs and gateway nodes obtained in the sample graph of
   Fig. 15.16 by the *Highest Degree* clustering algorithm.

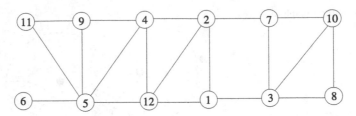

Fig. 15.16 Sample graph for Exercise 3

Fig. 15.17 Sample graph for
Exercise 4

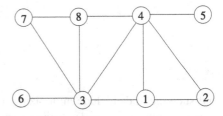

4. Implement *Wu_MCDS* algorithm in the sample graph of Fig. 15.17 by showing
   the nodes marked in the first phase and the nodes that are pruned in the second
   phase.

References

1. Alzoubi KM, Wan P-J, Frieder O (2002) New distributed algorithm for connected dominating
   set in wireless ad hoc networks. In: Proc 35th Hawaii international conference on system
   sciences, Big Island, Hawaii, 2002
2. Alzoubi KM, Wan P-J, Frieder O (2002) Message-optimal connected dominating sets in mobile ad hoc networks. In: MOBIHOC, EPFL Lausanne, Switzerland, 2002
3. Amis AD, Prakash R, Vuong THP, Huynh DT (2000) Max–min D-cluster formation in wireless ad hoc networks. In: Proc IEEE INFO-COM, Tel Aviv, Israel, 2000
4. Anupama M, Sathyanarayana B (2011) Survey of cluster based routing protocols in mobile ad
   hoc networks. Int J Comput Theory Eng 3(6):806–815
5. Banerjee S, Khuller S (2000) A clustering scheme for hierarchical routing in wireless networks. Technical report CS-TR-4103, University of Maryland, College Park
6. Basagni S (1999) Distributed clustering for ad hoc networks. In: Proc internat sympos parallel
   algorithms, architectures and networks ISPAN'99, Perth, Australia, pp 310–315
7. Basagni S (1999) Distributed and mobility-adaptive clustering for multimedia support in
   multi-hop wireless networks. In: Proc IEEE VTC, September 1999
8. Cidon I, Mokryn O (1998) Propagation and leader election in multihop broadcast environment.
   In: Proc 12th int symp distr computing, pp 104–119
9. Cokuslu D, Erciyes K, Dagdeviren O (2006) A dominating set based clustering algorithm
   for mobile ad hoc networks. In: International Conference on Computational Science, vol 1,
   pp 571–578
10. Das B, Bharghavan V (1997) Routing in ad-hoc networks using minimum connected dominat-
    ing sets. In: IEEE international conference on communications (ICC'97), vol 1, pp 376–380
11. Du DZ, Pardalos P (eds) (2004) Handbook of combinatorial optimization. Kluwer Academic,
    Dordrecht, pp 329–369

12. Erciyes K, Ozsoyeller D, Dagdeviren O (2008) Distributed algorithms to form cluster based spanning trees in wireless sensor networks. In: ICCS 2008. LNCS. Springer, Berlin, pp 519–528

13. Garey MR, Johnson DS (1979) Computers and intractability: a guide to the theory of NP-completeness. Freeman, New York

14. Gerla M, Tsai JTC (1995) Multicluster, mobile, multimedia radio network. Wirel Netw 1:255–265

15. Ghiasi S, Srivastava A, Yang X, Sarrafzadeh M (2002) Optimal energy aware clustering in sensor networks. Sensors 2(7):258–269

16. Guha S, Khuller S (1998) Approximation algorithms for connected dominating sets. Algorithmica 20(4):374–387

17. Kawadia V, Kumar PR (2003) Power control and clustering in ad hoc networks. In: Proc of IEEE INFOCOM 2003, vol 1, pp 459–469

18. Krishna P, Vaidya NH, Chatterjee M, Pradhan DK (1997) A cluster-based approach for routing in dynamic networks. Comput Commun Rev 49:49–64

19. Li L, Halpern JY, Bahl P, Wang Y, Wattenhofer R (2001) Analysis of cone-based distributed topology control algorithm for wireless multi-hop networks. In: Proc 20th annual ACM SIGACT-SIGOPS symposium on principles of distributed computing (PODC)

20. Li X-Y, Wan P-J, Wang Y, Frieder O (2002) Sparse power efficient topology for wireless networks. In: IEEE Hawaii international conference on system sciences (HICSS 2002)

21. Li L, Halpern JY, Bahl P, Wang Y-M, Wattenhofer R (2005) A conebased distributed topology-control algorithm for wireless multi-hop networks. IEEE/ACM Trans Netw 13(1):147–159

22. Lin CR, Gerla M (1997) Adaptive clustering for mobile wireless networks. IEEE J Sel Areas Commun 15(1):1265–1275

23. Liu JS, Lin CHR (2005) Energy-efficiency clustering protocol in wireless sensor networks. Ad Hoc Netw 3(3):371–388

24. Lukovszki T (1999) New results on geometric spanners and their applications. PhD thesis, University of Paderborn

25. Manousakis KS, Baras JS (2003) Clustering for transmission range control and connectivity assurance for self configured ad hoc networks. In: Proc of IEEE MILCOM 2003

26. Nocetti FG, Solano-González J, Stojmenovic I (2003) Connectivity based k-hop clustering in wireless networks. Telecommun Syst 22(1–4):205–220

27. Peleg D (2000) Distributed computing: a locality-sensitive approach. SIAM, Philadelphia. ISBN 0-89871-464-8

28. Rekhter Y, Li T (1995) A border gateway protocol 4. IETF RFC 1771

29. Wattenhofer W, Li L, Bahl P, Wang Y-M (2001) Distributed topology control for power efficient operation in multihop wireless ad hoc networks. In: Proc IEEE infocom

30. Wei D, Chan HA (2006) Clustering ad hoc networks: schemes and classifications. In: Sensor and ad hoc communications and networks (SECON 2006), vol 3, pp 920–926.

31. Yu J, Chong P (2005) A survey of clustering schemes for mobile ad hoc networks. IEEE Commun Surv Tutor 7(1):32–48

Chapter 16
Ad Hoc Routing

Abstract Routing is the process of deciding on the optimal route between a source node and destination node in a network. A routing protocol is a collection of algorithms and procedures to perform routing. We have already investigated distributed routing algorithms in wired computer networks. Our focus in this chapter is on the routing protocols rather than algorithms for wireless ad hoc networks, and we will see that the routing in such networks have different requirements due to the mobility and energy limitations of the nodes.

16.1 Introduction

Distance Vector Routing based on Bellman–Ford Algorithm and Link State Routing based on Dijkstra's Algorithm are two of the most important dynamic routing algorithms used in wired networks as discussed in Chap. 7. In distance vector protocols, nodes periodically broadcast their neighbor information and update their tables based on the information received. A known problem with these protocols is the slow convergence time as the information broadcast should be propagated to all nodes in the network, which also causes the *count-to-infinity* problem. In link state protocols, nodes broadcast their neighbor information to the network, and each node builds a complete communication graph of the network and then can compute the shortest routes using an algorithm like Dijkstra's algorithm. Both protocols calculate the distance between nodes using metrics such as hop count, capacity of link, or queue delay. Compared to link-state, the distance vector protocol is easier to implement and requires less storage space.

In a MANET, frequent changes in the topology of the network require the routing information to be broadcast in short intervals using the scarcely available bandwidth. Also, routing algorithms in ad hoc networks should consider limited energy, limited bandwidth, and high error rates. For this reason, the Distance Vector and Link State protocols require considerable modifications to be used in MANETs and WSNs.

16.2 Characteristics of Ad Hoc Routing Protocols

Although there are numerous routing protocols for ad hoc wireless networks, there is not a single protocol that meets the demands of all applications. An efficient

K. Erciyes, *Distributed Graph Algorithms for Computer Networks*,
Computer Communications and Networks, DOI 10.1007/978-1-4471-5173-9_16,
© Springer-Verlag London 2013

way to evaluate a protocol designed for ad hoc wireless networks is to analyze its
time and message complexities like we have done for most of the distributed algorithms. Implementing a protocol using a simulator such as the ones discussed
in Chap. 14 and evaluating its performance will also give a good idea on its performance for different traffic loads. There are different criteria for designing and
classifying routing protocols for wireless ad hoc networks such as the contents
of the routing information exchanged, when and how this information routing is
transferred, the computation of routes, etc., which are described in the next sections.

16.2.1 Proactive and Reactive Protocols

In *source initiated routing*, the source determines all the intermediate nodes that
the message has to pass through, prior to sending the message, and it inserts these
node identifiers whether identities or IP or MAC addresses on the path to destination in the message header. Any intermediate node on the path then simply
forwards the message to the next node in the list of the header. In *destination-
based* or *table-driven routing*, only the destination address is stored in the header,
and the nodes along the path determine the best next node to reach the destination using their current local information of the network stored in their routing tables.

In *proactive routing protocols*, routes are stored in local tables at nodes prior
to sending messages to destinations, and therefore these protocols are table-driven.
When the network topology changes due to the movement of some nodes as in a
MANET, update messages should be broadcast to provide a consistent view of the
network by the nodes. A disadvantage of these protocols is that the update messages
have significant overhead. Few sample proactive routing protocols will be described
in Sect. 16.3.1.

In *reactive routing protocols*, routes to destinations are constructed only when
needed, and therefore these protocols are source initiated. These protocols start
a route discovery process before sending a message to a destination. This process basically consists of sending packets with a description of the destination
(address information of the destination) between the nodes to discover the route.
Any node receiving such a request looks into its available routing table to find if
it has a route to the described destination. If a route to the destination is present,
the node returns this route to the source, and the process ends; else the request
packet is forwarded to the neighbors continuing the route search process. Once
a route is found, it is temporarily maintained in the routing table and then subsequently removed either after a timeout or if the destination node leaves the
network. Since there are no periodic or immediate updates due to the mobility
of the nodes, the control message overhead is less than proactive protocols resulting in better scalability. Few example reactive protocols will be reviewed in
Sect. 16.3.2.

16.2.1.1 Flat or Hierarchical Protocols

Routing in ad hoc wireless networks can be realized as flat or hierarchical. In flat routing structure, all nodes have the same routing functions. However, this method is not suitable for large networks consisting of hundreds of nodes. For such large networks, hierarchical routing where the networks is partitioned into a group of close nodes called clusters is usually preferred. Each cluster is managed by a special node, called clusterhead, which is responsible for any external communications of the cluster. This approach reduces the amount of control information exchange and has a better performance than the flat structure for large networks.

16.2.1.2 Adaptive or Non-adaptive Protocols

In nonadaptive routing protocols, the routes are calculated in advance and sent to the routers when the networks is initialized. In adaptive protocols, the routes are recalculated when the network topology or traffic changes. Adaptive protocols are dynamic, and MANETs use only this type of protocols as their topology is dynamic.

16.3 Routing in Mobile Ad Hoc Networks

The basic requirements from any routing protocol in a MANET are that it should consider the mobility and limited energy of the nodes as well as limited bandwidth and high error rates of the wireless communication medium. The protocols should be distributed, and the routes found should be loop-free.

In order to provide uninterrupted communication in a MANET, discovering routes dynamically can be achieved either periodically or event based. In general, implementation of protocols based on periodic updates is simpler and provides more stable networks. However, they result in significant control messages using the scarce bandwidth of the wireless network. Increasing the period duration reduces the number of control messages; however, the network topology stored may not reflect the existing structure.

In event-based protocols, the network information should be sent only when there is a topology change such as a link failure due to the movement of a node or the forming of a new link as a new node enters the area. For a highly dynamic network where nodes move rapidly, event-based protocols require significant number of control messages. Two classes of protocols for MANETs that reflect these design issues are proactive and reactive protocols, examples of which are described next.

16.3.1 Proactive Protocols

We will investigate two important protocols, called DSDV and WRP, that may be used in a MANET in this section.

16.3.1.1 DSDV Protocol

The Destination-Sequenced Distance-Vector Routing protocol (DSDV) is a proactive protocol based on the Bellman–Ford algorithm with some improvements [12]. It aims to provide routing in ad hoc networks based on distance vector protocols by preventing the count-to-infinity problem using sequence numbers.

In DSDV, each node in the network has a routing table for all destinations where each entry shows the destination, number of hops to reach there, and a sequence number. Each node periodically broadcasts its routing table and also when significant new routing information becomes available. The broadcast packet contains destination address, the number of hops required to reach the destination, and the sequence number received related to the destination for each new table entry. The sequence number and hardware address of the transmitter node are also included. Any node that receives the information packet increments the hop count and broadcasts it to enable the reception of the packet by all nodes.

The communication links may be broken as nodes move, and this may be detected by the Layer 2 (Data Link Layer) protocol, or it may be found if there are no broadcasts for a duration of time from a neighbor. A broken link detected by a neighbor is assigned the parameter ∞, and any route through this link is assigned an odd sequence number and ∞ as shown in Fig. 16.1, where node f that comes in the vicinity of node c broadcasts $\langle f, 27, 0 \rangle$, which means that it has 0 hops to f (itself) with a sequence number 27. Node c receives this message, increments hop count, and broadcasts $\langle f, 27, 1 \rangle$ to inform neighbors that node f can be reached by itself by one hop. The other nodes update their tables similarly. In (b), the link between nodes c and d is broken as node d moves away. Node c discovers the broken link and informs neighbors by the $\langle d, 45, \infty \rangle$ message, meaning that d cannot be reached via node c anymore. All other nodes update their tables accordingly.

Routing tables in DSDV are updated periodically using two types of messages. In *full dumps*, several network protocol data unit (NPDU) packets are transmitted to transfer all routing table information. In *incremental updates*, a single NPDU is used to send only the information that has changed since the last full dump. When a node receives information about a new route, this is compared with previous information packets for the same route and a route with a newer sequence number is used to update the tables.

Disadvantages of DSDV include frequent updating of the routing tables and for a highly dynamic network, a high convergence time. For these reasons, DSDV is not much used, but it provides a basis for other protocols.

16.3.1.2 WRP Protocol

Wireless Routing Protocol (WRP) is a proactive unicast routing protocol based on distance vector routing for mobile ad hoc networks (MANETs) [9]. It is similar to DSDV so that each node in the network has tables showing routes to all destinations. However, the table storage and table updating procedures are different than DSDV. WRP stores a number of tables at each node as follows:

Fig. 16.1 DSDV example

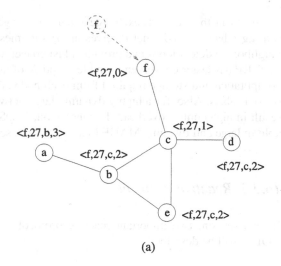

(a)

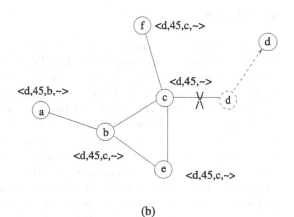

(b)

- Distance Table (DT): This table has an entry for each neighbor of a node, showing the distance and the predecessor node for a particular destination as informed by the neighbor node.
- Routing Table (RT): This table provides the current routes for all destinations from the node. For each entry, it keeps the shortest distance, the predecessor node, the successor node, and a flag. The flag shows whether the path is simple, a loop (error), or not a valid one. Storing predecessor and successor nodes is used to prevent loops.
- Link Cost Table (LCT): It shows the cost, as the number of hops, of transferring messages over each link.
- Message Retransmission List: It has an entry for each update message to be re-transmitted. Whenever an update message is transmitted, each entry in routing table is marked to wait for an acknowledgement. The number of acknowledgements not received indicates link failures.

Nodes in the network exchange *update* messages periodically to update their routing tables. A node that receives an *update* message also tests its routes with neighbors to detect loops and provide a fast convergence of routes.

WRP has faster convergence and less update of tables than DSDV. However, the computation and storage required from each node is higher due to the maintenance of four tables. Also, for a highly dynamic large network, frequent update messages result in high control overhead. In conclusion, as DSDV, WRP protocol is not suitable in large and dynamic MANETs as it does not scale well.

16.3.2   Reactive Protocols

In this section, two important reactive protocols for MANETs, called DSR and AODV, will be described.

16.3.2.1   DSR

The Dynamic Source Routing (DSR) protocol was developed at Carnegie Mellon university [6] and was designed for MANETs. The motivation behind DSR was to create a routing protocol that had very low overhead and also could react quickly to changes in the network due to mobility of nodes. It is based on source routing in which a list of the nodes that the packet must pass through is provided by the source node in the packet. This method eases the task of the intermediate nodes as they do not need to store any routing information; however, they may store the routing information in tables to improve performance. As the networks and their diameters get larger, the amount of information included in the packet header increases, and for this reason, DSR can be used from small- to medium-sized networks with favorable performance but is not advantageous in large diameter networks. The designers of DSR assumed that the diameter of the network is not greater than 10 hops and the mobile nodes may move at any time but with moderate speeds [6]. It has two main phases, the *route discovery* and *route maintenance*. Both route discovery and route maintenance are activated on demand without the need for periodic update packets, which would mean that there will be no routing packets when the nodes are stationary relative to each other.

Route Discovery

Route discovery is the procedure to obtain a route to a destination and is only activated by the source when this information is not available. When a source node u wants to send a packet to a destination node v, it checks its cache. If it does not have any prior knowledge of the route, it sends a *route_request(source, destination, id)* (*RREQ*) message to its current neighbors. An intermediate node receiving this packet has three options as below:

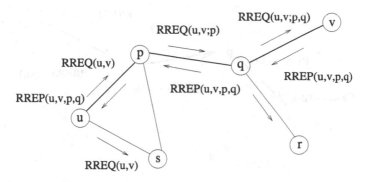

Fig. 16.2 DSR example

1. If a packet was received before, it discards it as this is a duplicate packet.
2. Else it checks its *route cache* to find if a route to destination is already known. In this case, it replies by the *route_reply* message to the sending node.
3. Else it adds itself to the route list in the packet, and it broadcasts RREQ.

When the RREQ packet reaches the destination node v, it copies the accumulated route information in RREQ to RREP and sends it to the source along the path that the packet has traversed reaching v. The packet now has all the intermediate node identifiers as shown in Fig. 16.2. When the source u receives RREP, it stores the routing information in its route cache for further use. For unidirectional routes, also node v has to initiate a route discovery procedure to u.

Route Maintenance

Route maintenance of DSR provides procedures for link breakdown, route reply storms, and limiting the number of hops. Any node that notices the breakdown of a link sends a route error message (*RERR*) to the source. Any node that detects RERR message updates its route cache to remove the link. Route reply storms are possible when several nodes respond by RREP messages to the RREQ message by the source as they have the route in their cache and these replies may collide. To prevent such a situation, nodes may send replies to the source by random delays. Each RREQ message contains a hop count which is an integer decremented by each node along the path to the destination, and if 0 is reached, the RREQ message is discarded.

The advantages of DSR are that the routes are maintained only in the source and destination nodes, and the resource discovery phase may provide multiple paths to the destination. The disadvantages are that flooding becomes a problem with nodes moving fast and that the packet size grows significantly larger with growing network diameter.

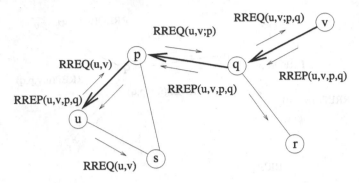

Fig. 16.3 AODV example

AODV

Ad hoc On Demand Distance Vector (AODV) is a reactive MANET protocol, which is similar to DSDV, aiming to decrease the network control traffic as much as possible [13]. AODV also provides unicast and multicast communications and has a routing table for each. When a route in AODV is not used for a period of time, it is discarded reducing the route maintenance overhead.

In AODV, routes are discovered only when needed, as in other reactive protocols. Node u wishing to send a message to a destination node v checks its table whether such a route exists and initiates a route discovery process by the Route Request (RREQ) message if there is not a route. RREQ message contains IP addresses and the current sequence numbers of the source and destination nodes and also a broadcast identifier number. Any intermediate node that receives RREQ message sets up a reverse link to point to the sender of this message and floods the neighbors with RREQ message, which reaches the node v if the network is connected. The destination v replies with a unicast Route Reply (RREP) packet if the IP in the incoming RREQ packet message matches its IP and the sequence number in the packet is at least as large as its sequence number to prevent loops, an which traverses the intermediate nodes through the reverse links to the source node u at which point a route between u and v is established as shown in Fig. 16.3. Data packets (DATA) that do not have any routing information can now follow the path taken by the RREQ packets to reach the destination.

If an intermediate node w has already a route to the destination v, it replies to the RREQ packet with RREP packet, and when this packet reaches the source node u, it updates its table so that w is included in the table. An intermediate node may receive more than one RREP packet and forwards the first one it receives unless the RREP packets it later receives have a higher sequence number or a smaller hop count.

Whenever the source node moves, a new route discovery process needs to be started. If intermediate nodes or the destination node moves, links may be broken. In this case, routing tables are updated, and the Route Error (RERR) messages are sent to active neighbors. Neighbor nodes exchange HELLO messages periodically to check each other, and absence of these messages indicate link failure.

TORA

The Temporally Ordered Routing Algorithm (TORA) is a flat, reactive routing protocol based on the link reversal concept, designed for highly dynamic MANETs [10, 11]. The general idea of this protocol is to build and maintain a directed acyclic graph (DAG) routed at the destination. There may be more than one DAG, and therefore multiple paths to the destination may exist. The height data structure used in TORA is a quintuple of the form $h = (t, oid, r, d, i)$, where t is the time, oid is the originating node identity, r is the reflecting bit, d is the ordering integer, and i is the identity of the node. Each node in TORA has a table showing the heights of the neighbors as quintuples in terms of two parameters. The first is the *reference level* as indicated by the three parameters, and the second is the *offset* from the reference level shown by the last two parameters. Node i initially has a NULL height as $(-, -, -, -, i)$, and a destination node v has $(0, 0, 0, 0, v)$ as its height. Information flows from higher height nodes to lower height nodes toward the destination. TORA maintains a totally ordered set of heights that provides a loop-free operation.

There are three operations in TORA: route creation, route maintenance, and route erasure. Node u wishing to send data to a destination node v, which does not have any downstream nodes for v, broadcasts a *query* (QRY) packet to its neighbors and sets its Route Request (RR) flag. A node that receives QRY packet does the following:

1. If RR is 1 (already has an outstanding query), it discards QRY packet.
2. If RR is 0, and no downstream links for the destination exists, it broadcasts QRY packet.

The QRY packet is flooded through the network in this manner until it reaches the destination node v or an intermediate node w that has a route to the destination. The reply to the QRY packet is the *update* (UPD) packet. The destination node v or the intermediate node w sends a UPD packet to inform neighbors that the destination is discovered. A node that receives a UPD packet does the following:

1. If RR is 1 and the reflection bit of the neighbor height is zero, it increments the value of its neighbor's height in the packet and stores this as its height. It then sends the UPD message including its height to its neighbors.
2. If RR is 0 and reflection bit of the neighbor is set, meaning that the route of the neighbors is invalid, it only updates neighbor's entry in the table.

At the end of this process, there may be a number of routes from the source node u to the destination node v and u may choose the shortest route. A node may discover the loss of a downstream link due to the movement of nodes. In this case, the discovering node creates a new reference level and broadcasts this reference level to its neighbors, resulting in the reversing of the links of the DAG. This way, control messages are confined to a small region where the link is lost. When a node detects a network partition, it generates a clear (CLR) packet, which is flooded through the network, resulting in erasing of the invalid routes in the network.

Advantages of TORA are the provision of loop-free operation, the existence of multiple paths to the destination that provides fault tolerant routing, and relatively local updates in case of route changes. An important disadvantage is that the routes may be far from optimal due to local reconfiguration of paths.

16.3.3 Hybrid Routing Protocols

Hybrid routing protocols have the advantages of both proactive and on-demand routing methods, and they are used in networks of large size. In the following, we will examine such a protocol, called the *Zone Routing Protocol*.

16.3.3.1 The Zone Routing Protocol

In this protocol, each node has a zone around itself defined by the number of hops defined as the *zone radius* [2]. For nodes within this zone, a node uses proactive routing to communicate, and for other nodes outside the zone, the on-demand routing strategy is used. The protocol has three main modules. For nodes within the zone, the Intra-zone Protocol (IARP) is used, which can be a link state or a distance vector protocol. On the other hand, the Inter-zone Protocol (IERP) provides routing for nodes outside the zone using the on-demand routing method. IERP has Route Request (RREQ) and Route Reply (RREP) packets similar to other on-demand protocols. IERP uses the Bordercast Resolution Protocol (BRP) for the border nodes to broadcast queries for destinations not in the zone. Broadcasting of the RREQ packets is performed from a node border to the borders of other nodes until the destination node is found within a zone. This protocol has the advantage of keeping the overhead of IARP within the zone; however, flooding of the RREQ packets is still a problem.

16.4 Routing in Sensor Networks

Routing in a WSN has different challenges than in a MANET for the following reasons. First, any routing scheme based on identifiers of nodes is difficult to apply as maintaining identifiers in a large sensor network consisting of hundreds of nodes is difficult. In general, the identifiers of nodes are not needed when collecting data from sensor nodes. Second, the flow of data has an orientation, mostly from sensor nodes to the data gathering point called the sink or the base station, rather than arbitrary data exchange between individual nodes. Third, the sensor nodes are stationary in most cases with only few of them being mobile resulting in simpler route maintenance procedures. Fourth, sensor nodes are application specific, and data collection is usually based on the location. Also, the energy in a sensor network is limited, and data has some redundancy as it is collected by many nodes that are in the same location.

Routing protocols for sensor networks should consider these properties. For example, data redundancy should be considered by the routing protocols for more efficient communications. Special techniques such as data aggregation, clustering, and data centric operations are frequently used. We will classify routing protocols for sensor networks as data-centric, hierarchical, and location-based routing protocols and give examples of these methods in the next sections.

16.4.1 Data-Centric Protocols

In order to overcome the data redundancy problem, routing protocols that select a group of sensors and aggregate this data are designed. This type of protocols are called *data centric*, where the sink sends queries to certain regions of the network and collects data from these regions. The queries sent by the sink use attribute-based naming to specify data properties. For example, the sink may request data related to temperature of an environment as $\langle T \geq 30 \rangle$, meaning that only data sensed as greater than 30 °C is required. The first data-centric protocol is called SPIN and later Directed Diffusion is developed, as described next.

16.4.1.1 Flooding and Gossiping

A simple approach for transferring data in a sensor network is to provide the transfer without routing. Flooding is such a method, and similarly to the flooding, we have seen in Chap. 4 that any node that receives data simply broadcasts it to all neighbors, and this process is repeated until all neighbors receive data or some predefined hop count is reached. Duplicate packets are a problem in flooding, and *gossiping protocols* provide improvement over flooding by randomly selecting a neighbor node to send data [3]. However, although gossiping reduces the number of messages, it may introduce further delays.

16.4.1.2 SPIN

Sensor Protocols for Information via Negotiation (SPIN) is a family of protocols that names data using high-level descriptors. In SPIN, a node that receives new data advertises this to its neighbors that are interested, and they can receive this data by requesting. Three types of messages used in SPIN are *advertise* (ADV), *request* (REQ) to request specific data, and DATA messages to transfer data.

Figure 16.4 displays a sensor network that uses SPIN where node 4 has some data available and advertises this by ADV message to neighbors 5 and 7, which request the data by the REQ message as shown in (a). Data is transferred to these

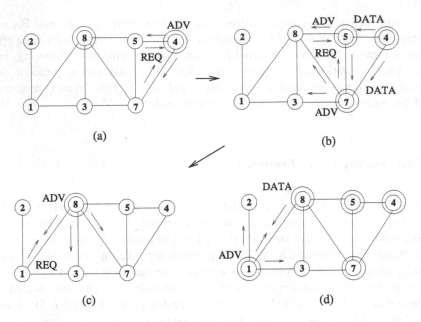

Fig. 16.4 SPIN example

nodes in (b), which in turn send ADV messages to their neighbors. Node 3 is not interested, and node 8 responds by an REQ message. In (c), data is transferred to node 8, which in turn advertises it to neighbors 1 and 3. Node 1 is interested and replies by an REQ message. Data is transferred to node 1 in (c), which can now send ADV to neighbors.

SPIN reduces the energy dissipation for a factor of 3.5 with respect to flooding and meta-data negotiation almost halves the redundant data [1]. However, if there exists a group of uninterested nodes between the nodes that request data and the nodes that have data, then the delivery of data to the interested nodes is prevented. For this reason, SPIN may not be used in critical sensor network applications such as military applications or intruder detection where reliability is important.

16.4.1.3 Directed Diffusion

Directed diffusion uses a naming procedure for data to transfer it to the requesting node [5]. It uses attribute-value pairs of data to query the nodes that have data. Data is sent in response to a query by the sink, whereas in SPIN, sensor nodes that have available data advertise to inform any potential receivers. In directed diffusion, nodes can perform data aggregation easing the task of the sink. However, this type of routing is not suitable for sensor network applications such as habitat monitoring, where continuous data delivery to the sink may be required.

16.4.2 Hierarchical Protocols

For ad hoc wireless networks of large size, the flat routing protocols result in high communication costs. Hierarchical routing protocols provide efficient scalable routing with decreased link and processing overheads in MANETs and WSNs. In this type of routing, nodes are organized into groups, called clusters, each with a cluster-head (CH). A CH is responsible for intercluster routing, and nodes within a cluster communicate with the CH only.

There may be two or more layers of clusters where the CHs of a lower layer are connected to the CHs of higher layer. In a sensor network, data is transferred from lower-level clusters to higher-level clusters toward the sink where each CH performs data aggregation and routing. CHs are chosen from the nodes with higher energy, and lower-energy nodes are involved in sensing only. In this section, we will review sample hierarchical routing protocols in WSNs.

16.4.2.1 LEACH

Low-Energy Adaptive Clustering Hierarchy (LEACH) is a cluster-based hierarchical routing algorithm for sensor networks [4]. In LEACH, clusters are formed based on the received signal strength, and a CH for each cluster is chosen. CHs are responsible for routing decisions, and all nodes in a cluster communicate directly to the CH. CHs are rotated to balance energy dissipations. A node becomes a CH with a probability for a round, and a new CH is elected for the next round.

16.4.2.2 TEEN and APTEEN

Threshold sensitive Energy Efficient sensor Network protocol (TEEN) [7] combines data-centric routing with the hierarchical structure. It forms two levels of clusters in a sensor network by forming second-level clusters around the sink and the first-level clusters around the second-level clusters as shown in Fig. 16.5. TEEN is designed to provide timely response to sudden changes in sensed data. The CHs in TEEN send a hard and a soft threshold value to its members after the clusters are formed. These two values are used to limit the transmissions by comparing the value of sensed data against them; however, for routine monitoring, TEEN does not provide any periodic data.

The Adaptive Threshold sensitive Energy Efficient sensor Network protocol (APTEEN) [8] is an extension to TEEN and overcomes the problem with TEEN by providing periodic data to the application. The hierarchical cluster formation in APTEEN is similar to TEEN. There are three types of queries in APTEEN: historical, one-time, and persistent. The historical query analyzes previous sensed data values, one-time query results in the current state of data, and persistent query is used to monitor an event for a duration of time. In terms of energy dissipation and network lifetime, both TEEN and APTEEN have better performances than LEACH [1].

Fig. 16.5 TEEN example

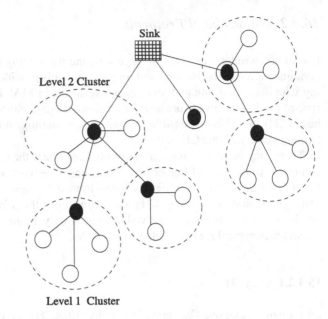

Level 2 Cluster

Level 1 Cluster

Fig. 16.6 GAF virtual grid

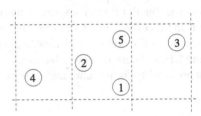

16.4.3 Location-Based Routing

Sensor nodes do not have identifiers in many applications. However, their location information can be used for routing purposes. In order to reduce transmissions, the query to obtain sensed data can be directed to a region. In this section, we will describe two location-based routing protocols, which may also be used for MANETs.

16.4.3.1 GAF

Geographic Adaptive Fidelity (GAF) [14, 15] is a location-based routing protocol that can be used in WSNs and MANETs. GAF forms a virtual grid in the network, and each node determines its location in this grid using GPS equipment. If two or more nodes are located in the same cell of the grid, they are considered equal, and all of these nodes except one can sleep. An example grid is shown in Fig. 16.6, where nodes 1, 2, 5 share the same grid cell, and any two of these nodes can be passive at any time saving considerable energy.

Fig. 16.7 States of a GAF
node

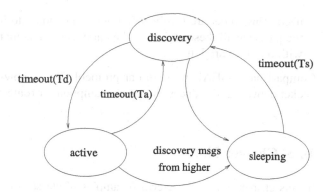

A sensor node in GAF can be in one of the three states as *sleeping, discovery*, and *active* as shown in Fig. 16.7. A node starts in *discovery* state and exchanges discovery messages containing grid and node identifiers, node state, and estimated node active time. The grid identity is computed using location information and the grid size. A timer T_d is set in this state, and when it expires, the node broadcasts its discovery message and enters the *active* state. In this state, another timer T_a is set, and when this fires, the node returns to the *discovery* state. A node in the *discovery* or *active* states can change to the *sleeping* state if another node using a ranking procedure is found, which can provide routing functions. A node in the *sleeping* state enters the *discovery* state after time T_s and starts checking its neighbors. The value of T_s can be adjusted to low values for MANETs with high mobility. GAF aims at keeping network connectivity by having an active node in each grid cell at all times.

16.4.3.2 GEAR

Geographic and Energy Aware Routing (GEAR) [14, 16] is a location-based routing protocol that uses location information when sending queries to certain regions of the network. Instead of broadcasting interests to whole network as in Directed Diffusion, GEAR sends queries to a defined region in the graph using *energy-aware* and *geographically informed neighbor* heuristics. Each node in GEAR stores an estimated cost and a learning cost of reaching a destination. The estimated cost is based on the residual energy of a node, its distance to the destination and the learned cost is a further refinement of the estimated cost. A packet is transferred to a region in two phases:

- Inter-region packet forwarding: When a node receives a packet, it forwards the packet to a neighbor that is closer to the region than itself. If there are no closer neighbors, a neighbor based on the estimated cost is chosen to forward the packet.
- Intra-region packet forwarding: When the packet reaches the region, it can be broadcast to all nodes in the region using *restricted flooding*. However, when the network consists of densely deployed sensors, *recursive geographic flooding* is

used. This procedure involves dividing the region into four regions and sending the packet to all these regions and recursively activating this routine until regions with one node are left.

Comparison of GEAR to a similar protocol GPSR shows that it performs better packet delivery and reduces energy consumption for route setup [16].

16.5 Chapter Notes

In this chapter, we have described samples of classes of the existing routing protocols for mobile and sensor networks. It was indicated that these two types of ad hoc wireless networks have different requirements; therefore, routing protocols for them have different goals in many cases. Routing protocols for mobile networks should consider mobility of nodes and interference problems, whereas mobility in general is not a primary concern in a sensor network that consists of mainly stationary nodes. However, nodes in general do not have identifiers in sensor networks, and methods using data attributes and location information are widely used in routing protocols for such networks. Hierarchical routing is used for both networks, and we have described few protocols for MANETs that use hierarchical routing. Other methods of clustering such as direct clustering and dominating sets were covered in Chap. 15.

Routing is a fundamental task in any computer network and will probably remain an active area of research for both wired and wireless networks, including MANETs and WSNs.

16.5.1 Exercises

1. Compare proactive and reactive routing methods in terms of when and how routes are established and the communication costs involved.
2. What are the main differences between the AODV and WRP protocols?
3. Compare DSR and AODV in terms of establishing routes between source and destination nodes.
4. Discuss how coordinates of sensors can be used effectively for routing in a WSN.
5. Describe an example of WSN routing protocol that uses clusters and CHs for routing.

References

1. Akkaya K, Younis M (2005) A survey on routing protocols for wireless sensor networks. Ad Hoc Netw 3:325–349

2. Haas ZJ, Pearlman MR (1998) The Zone Routing Protocol (ZRP) for ad hoc networks. IETF Internet draft
3. Hedetniemi S, Liestman A (1988) A survey of gossiping and broadcasting in communication networks. Networks 18(4):319–349
4. Heinzelman W, Chandrakasan A, Balakrishnan H (2000) Energy-efficient communication protocol for wireless sensor networks. In: Proc Hawaii international conference system sciences
5. Intanagonwiwat C, Govindan R, Estrin D (2000) Directed diffusion: a scalable and robust communication paradigm for sensor networks. In: Proc 6th annual ACM/IEEE international conference on mobile computing and networking (MobiCom'00)
6. Johnson DB, Maltz DA (1996) Dynamic source routing in ad hoc wireless networks. In: Imielinski T, Korth H (eds) Mobile computing. Kluwer Academic, Dordrecht, pp 153–181. Chapter 5
7. Manjeshwar A, Agrawal DP (2001) TEEN: a protocol for enhanced efficiency in wireless sensor networks. In: Proc 1st international workshop on parallel and distributed computing, issues in wireless networks and mobile computing
8. Manjeshwar A, Agrawal DP (2002) APTEEN: a hybrid protocol for efficient routing and comprehensive information retrieval in wireless sensor networks. In: Proc 2nd international workshop on parallel and distributed computing, issues in wireless networks and mobile computing
9. Murthy S, Garcia-Luna-Aceves JJ (1996) An efficient routing protocol for wireless networks. Mob Netw Appl 1(2):183–197. Special issue on routing in mobile communication networks
10. Park VD, Corson MS (1997) A highly adaptive distributed routing algorithm for mobile wireless networks. In: Proc INFOCOM '97, sixteenth annual joint conference of the IEEE computer and communications societies. Driving the information revolution. Proceedings IEEE, vol 3, pp 1405–1413
11. Park VD, Corson S (1997) Temporally-ordered routing algorithm (TORA), version 1, functional specification. IETF Internet draft
12. Perkins CE, Bhagwat P (1994) Highly dynamic destination-sequenced distance-vector routing (DSDV) for mobile computers. Comput Commun Rev 24(4):234–244
13. Perkins CE, Royer EM (1999) Ad hoc on demand distance vector routing, mobile computing systems and applications. In: Proc WMCSA '99. Second IEEE workshop, pp 90–100
14. Roychowdhury S, Patra C (2010) Geographic adaptive fidelity and geographic energy aware routing in ad hoc routing. Int J Comput Commun Technol 1(2, 3, 4):309–313. Special issue
15. Xu Y, Heidemann J, Estrin D (2001) Geography-informed energy conservation for ad hoc routing. In: Proc 7th annual ACM/IEEE international conference on mobile computing and networking (MobiCom'01)
16. Yu Y, Estrin D, Govindan R (2001) Geographical and energy-aware routing: a recursive data dissemination protocol for wireless sensor networks. Technical report, UCLA-CSD TR-01-0023, UCLA Computer Science Department

Chapter 17
Sensor Network Applications

Abstract *Localization* is the method of providing the coordinates of sensors in 2-D plane so that these coordinates may be attributed to the sensed data to make it more meaningful and also network protocols such as routing may use this information. An important application of sensor networks is the *tracking* of mobile objects in the area of deployment to determine their trajectory. In this chapter, we first investigate methods to solve the localization problem and then describe few algorithms to track objects efficiently in sensor networks where distributed graph algorithms such as clustering and tree construction can be used for real-life applications.

17.1 Localization

The aim of localization is to find the location of a sensor or an object either as absolute coordinates or as relative coordinates to other sensors. In general, relative coordinate finding is easier to implement, and we will investigate this possibility.

A direct way of localizing sensors is to have all of them equipped with Global Positioning System (GPS) receivers, but this method is hardly used due to the costs involved. It may however be necessary to have a subset of sensor nodes with fixed locations to be equipped with GPS receivers, which can find their coordinates directly. These nodes are called *anchors* or *beacons* and serve as reference points for other ordinary nodes to find their coordinates. An ordinary node that is made aware of its location during initialization by preprogramming can also serve as an anchor node. The physical location of anchors have an important effect on the accuracy of the localization. It has been suggested that if the anchors are placed in a convex hull around the network, the positioning accuracy improves [3].

Quality parameters for a localization algorithm are its accuracy and precision. The *positioning accuracy* is the maximum deviation from the real position of the object. The precision is the averaged ratio of reaching the accuracy. Localization can be performed using either *range-based* or *range-free* methods, which are described next.

K. Erciyes, *Distributed Graph Algorithms for Computer Networks*,
Computer Communications and Networks, DOI 10.1007/978-1-4471-5173-9_17,
© Springer-Verlag London 2013

17.1.1 Range-Based Localization

Range-based localization methods rely on measuring the distance to the object under consideration. In the case of a sensor node, the strength of the transmission signal from this node can be used. For other objects in the sensor field, distances can be estimated based on the sensed data measurements of the object. In this section, we will review basic methods to estimate the distances of objects or sensors to their neighbors.

17.1.1.1 Received Signal Strength Indicator

As every sensor node has a radio for communication, the first possible approach is to use the radio to estimate distances. Every sensor has the facility to detect the power of a radio signal it receives, and using this information, it can estimate the distance of the source. This method of finding the distances is called *Received Signal Strength Indicator* (*RSSI*) method. Radio signals diminish with the square of the distance from the source they are transmitted as follows:

$$P_{\text{rec}} = k \frac{P_{tx}}{d^2},$$ (17.1)

where k is a constant related to the medium, and P_{rec} and P_{tx} are the power levels of the received and transmitted signals. Therefore,

$$d = \sqrt{\frac{k \cdot P_{tx}}{P_{\text{rec}}}}.$$ (17.2)

Assuming that P_{tx} is same for every sensor and known, a sensor can estimate its distance to a source using the above equation. However, RSSI introduces large errors as radio propagation is significantly affected by the environment. Recently, it was shown that better calibration of radio transmitters may decrease these errors significantly [13]. A clear advantage of RSSI-based distance computation is that it does not require any additional hardware.

17.1.1.2 Time Difference of Arrival

Time Difference of Arrival (*TDoA*) method is frequently used to estimate the distance of a node due to its high accuracy. In TDoA, each sensor node has a speaker and a microphone. The transmitting node u first sends a radio signal to the receiving node v, and after some delay t_d, it sends an audio signal to the node v. The node v records the time it receives radio (t_r) and the time it receives sound (t_s) signals. As radio signal propagates much faster than the sound, the difference of arrival of

Fig. 17.1 Time Difference of Arrival example

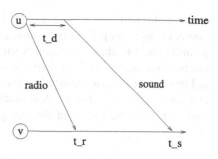

these two signals can be used so that node v (or any other listening node) can now calculate its distance d to u as follows:

$$d = (s_r - s_s)(t_s - t_r - t_d), \qquad (17.3)$$

where s_r and s_s are the propagation speeds of radio and sound waves, respectively. Fig. 17.1 shows an example of TDoA method to estimate the distance between nodes u and v. TDoA systems require the nodes to be equipped with speakers and microphones, which should be calibrated accurately. Also, the speed of sound varies with temperature and humidity, which causes decreased accuracy. However, due to the accuracy obtained, TDoA is a widely used method to estimate the distance of a node.

17.1.1.3 Angle of Arrival

In the *Angle of Arrival (AoA)* method, a node that is listening to a radio propagation tries to estimate the angle of arrival of a signal by computing time or phase analysis of the incoming signals received by the several antennas or microphones it has.

The AoA method has a high accuracy; however. it requires a more expensive equipment than TDoA. Also, accommodation of few antennas or microphones on a sensor board is difficult physically, and for these reasons, this method is hardly used in general applications.

17.1.2 Range-Free Localization

Range-free localization methods do not require the distance to the object to be estimated, and therefore they do not need special hardware but produce less accurate location information in general. One possible approach is to place anchors to form a regular mesh and to have them send signals of their coordinates periodically. The receiving sensors may then localize themselves to the connecting region of the anchors, which is the *centroid* of the anchors from which signals have been received [4]. Weights can be assigned to anchors based on their proximity to the sensor nodes and sensor nodes can find their positions as a weighted centroid of the connected anchors [11]. This simple and cost effective centroid method improves accuracy, but the choice of weights should be done according to some heuristic. In another approach, called *DV-hop*, the anchors transmit their coordinates as before, and the receiving nodes transmit them again until all nodes have these coordinates. The nodes can then calculate their distances based on the anchor locations, the hop counts from the anchors, and the average distance per hop [12].

17.1.3 Localization with Range Estimate

This type of localization has a number of anchors that obtain distances to the unknown node or object u by methods such as RSSI, TDoA, or AoA and cooperate to determine the coordinates of u. *Multilateration* uses simple geometry to calculate distances. In a two-dimensional plane, the distance between two nodes i and j using the Pythagoras theorem is given as follows:

$$d_{ij} = \sqrt{(x_i - x_j)^2 + (y_i - y_j)^2}.\tag{17.4}$$

For three anchor nodes a, b, and c and an unknown node u, this becomes

$$d_{au}^2 = (x_a - x_u)^2 + (y_a - y_u)^2,\tag{17.5}$$

$$d_{bu}^2 = (x_b - x_u)^2 + (y_b - y_u)^2,\tag{17.6}$$

$$d_{cu}^2 = (x_c - x_u)^2 + (y_c - y_u)^2,\tag{17.7}$$

where d_{au}, d_{bu}, and d_{cu} are the distances of nodes a, b, and c to the object u. If Eq. (17.7) is subtracted from Eqs. (17.6) and (17.5) to eliminate x_u^2 and y_u^2, the following system of equations is obtained:

$$\begin{bmatrix} x_c - x_a & y_c - y_a \\ x_c - x_b & y_c - y_b \end{bmatrix} \begin{bmatrix} x_u \\ y_u \end{bmatrix} = \begin{bmatrix} (d_{au}^2 - d_{cu}^2) - (x_a^2 - x_c^2) - (y_a^2 - y_c^2) \\ (d_{bu}^2 - d_{cu}^2) - (x_b^2 - x_c^2) - (y_b^2 - y_c^2) \end{bmatrix},\tag{17.8}$$

where substituting anchor coordinates yields the coordinates of the object u. Figure 17.2 displays an example network that has three anchors with coordinates as $a(2, 5)$, $b(9, 1)$, and $c(8, 6)$. The object u is detected by all three, and the distances to u are determined as $\sqrt{10}$, 5, and $\sqrt{13}$, respectively. Substituting these values into Eq. (17.8) yields

$$2 \begin{bmatrix} 6 & 1 \\ -1 & 5 \end{bmatrix} \begin{bmatrix} x_u \\ y_u \end{bmatrix} = \begin{bmatrix} 68 \\ 30 \end{bmatrix}.\tag{17.9}$$

Solving for x_u and y_u in Eq. (17.9) yields $6x_u + y_u = 34$ and $-x_u + 5y_u = 15$, resulting in $x_u = 5$ and $y_u = 4$, which are the actual coordinates of u as shown in Fig. 17.2.

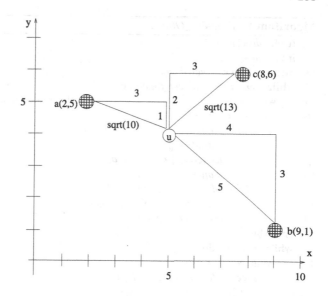

Fig. 17.2 Three-lateration example with three anchors

An implementation of the modified triangulation method is shown in Algorithm 17.1 (*Detect_Object*), where a node detecting an object waits for two neighbors to detect the same object, and based on the distance and coordinate values received from them, it applies Eq. (17.8) to find the coordinates of the object u. Any node that detects an object broadcasts its estimate of the distance to that object along with its coordinates by the *detected*(x, y, d) message, where x, y are the coordinates, and d is the estimated distance. When a node involved in triangulation computes the coordinates of an object, it broadcasts this information by the *coord* message to its neighbors. Nodes wait for detection or a *detected* message initially. We have assumed that nodes only know their own coordinates and a node that has not detected an object is not interested in computing the coordinates of that object. It however records the coordinates that it has received in anticipation of detecting the object.

17.2 Target Tracking

Target tracking is a fundamental WSN application where the presence of a mobile object such as an animal, vehicle, or an intruder is detected and its trajectory in the region can be estimated and monitored. Target tracking involves detection of the object by the nearby sensors, collaboration by them to determine the target location more precisely, and aggregation of the data to the sink for the determination of the trajectory. The *track quality* parameter is important to determine the quality of the tracking method used and is defined as follows.

Definition 17.1 (Track quality) The *track quality* is the maximum distance between the real track $\tau[i]$ of the target object and the approximate track $\tau'[i]$ obtained by

Algorithm 17.1 *Detect_Object*

1: **tuples** *dists*[] ← ∅
2: **int** *count* ← 1
3: **message types** *detected*, *coords*
4: **while** *event_type* ≠ *object_found* **do**
5: **wait** for event   ▷ either node or a neighbor detects an object
6: **case** *event_type* **of**
7:   *object_found*:   ▷ an object is detected
8:    **estimate** d_{iu}
9:    $dists[count] \leftarrow (x_i, y_i, d_{iu})$
10:    $count \leftarrow count + 1$
11:   $detected(x_j, y_j, d_{ju})$ received   ▷ another detection
12:    $dists[count] \leftarrow (x_j, y_j, d_{ju})$
13:    $count \leftarrow count + 1$
14: **end while**
15: **while** ¬*timeout* **do**
16: **while** *count* < 3 **do**
17:   **receive** $detected(x_j, y_j, d_{ju})$
18:   $dists[count] \leftarrow (x_j, y_j, d_{ju})$
19:   $count \leftarrow count + 1$
20: **end while**
21: **end while**
22: **find** x_u, y_u from *dists* using Eq. (17.8)
23: **broadcast** $coord(x_u, y_u)$ ▷ broadcast coordinates of object

---

the tracking algorithm as follows:

$$QT_{\text{track}} = \max_i \{\tau[i] - \tau'[i]\}. \tag{17.10}$$

In some specific target tracking applications such as boarder surveillance systems and military applications, *target classification*, where different procedures are needed to determine whether the target is an animal, vehicle, or an intruder, is also important. In general, target tracking algorithms can be classified as cluster-based, tree-based, and prediction-based as follows.

17.2.1 Cluster-Based Approaches

Cluster-based target tracking can be performed either by using static clusters or dynamic clusters as described below.

17.2.1.1 Static Clustering

In static cluster-based target tracking, clusters are formed statically at the time of network deployment. The size of a cluster, the area it covers, and the members it

Fig. 17.3 Static cluster-based tracking example

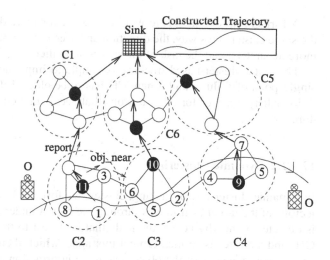

has do not change over the lifetime of the network. When an object moves within a cluster, some nodes will sense it and be activated. These nodes send their estimated location of the target to the clusterhead (CH) which can then calculate a more accurate position of the object along with its speed and estimated trajectory using previous data about the object and send this data to the sink.

Figure 17.3 displays a WSN that has been divided into six clusters $C_1, \ldots, C_6$ during initialization. There is also an overlay spanning tree T that connects all the nodes to the sink. The formation of clusters can be done independently from the formation of T, or an algorithm such as the one described in Sect. 15.4.8 can be used to provide clusters and a spanning tree at the same time.

An object O enters the region through C_2 and is detected by the nearby nodes 8, 1, and 3, which send their estimated distances to O to node 11, which is the CH of C_2. Node 11 then computes the location of O using the triangulation method and sends the estimated coordinates and other information such as the speed and the direction of O to the sink via the spanning tree T in a *report* message if data from the previous cluster is available. Similarly, when O enters C_3, nodes 6, 5, and 2 are activated, send their data to node 10, which then sends its computed coordinates of O to the sink via T. The sink can merge data received from the activated clusters C_2, C_3, and C_4 and can construct the trajectory of the moving object as shown. Static cluster-based tracking algorithm can be implemented similar to Algorithm 17.1 with the difference that multilateration is performed by the CH only.

Some improvements can be done to provide a more power efficient method with better estimate of the trajectory. The nodes of a cluster may be put to sleep, and they are only awaken when an *obj_near* message from a nearby cluster arrives. For example, the activated CH may send an *obj_near* message toward the next cluster along the trajectory of the object so that the nodes of the next possible cluster are awaken. For the example of Fig. 17.3, CH 11 may send this message to node 3 as it can estimate that the object will be moving toward C_3 and node 3 is the gateway node to C_3.

A further improvement is to wait for a while for more detection of the object in the same cluster. This way, the direction and speed of the object may be determined more accurately, and the task of the sink is simplified.

The static cluster-based tracking is simple to implement; however, a CH is a single point of failure, and a new CH has to be elected if it fails to function due to its battery drain or for other reasons that necessitate cluster maintenance procedures.

17.2.1.2 Dynamic Clustering

In dynamic cluster-based tracking, clusters are formed dynamically around the trajectory of the target as it moves through the region under control. When an object is detected by nearby nodes, the node that is closest to the object is elected as the CH, and a cluster is formed around this node, which then calculates the location, speed, and trajectory of the object using the information obtained from the neighbors as before. The newly formed CH may then inform the next node that is not in its cluster by the *obj_near* message, which then becomes the CH of the cluster that will be formed next along the trajectory of the object. However, since there is no fixed and maintained cluster structure, this approach is more fault tolerant as even if some nodes run out of battery power, the remaining nodes will continue to form clusters dynamically.

An example of dynamic cluster based tracking is shown in Fig. 17.4, where an object O enters the region, and nodes 2, 7, 3, and 11 detect it and form a cluster around it. After the election phase, node 2 is chosen as the CH as it has the lowest distance to O. The coordinates and the distance information are sent to this node, to estimate the coordinates of O, which can then be sent to the sink by a *report* message along the initially constructed spanning tree T as in the static cluster algorithm. As an improvement, nodes in this cluster may wait for a while in anticipation of further detections, and if this happens, the CH node 2 is informed which can assess the speed and direction of the object more accurately. Node 2 may then find its closest neighbor in that direction, which is node 3, and send an *obj_near* message that is transferred to node 9 to awaken. Node 9 can now awaken its neighbors in anticipation of O entering its region. Using this procedure, nodes may sleep most of the time saving energy, and only the nodes that are predicted to be on the movement direction of O are awaken.

Alternatively, the first nonmember node in the direction of the movement of the object that is awaken by the *obj_near* message may be assigned as the CH of the next cluster to be formed along the trajectory of the object as shown in Fig. 17.5.

We propose an algorithm to perform tracking using dynamic clusters, called *Dclus_TT*, assuming the following:

- Every node has a unique identifier.
- Each node knows its coordinates determined by any of the described localization methods, and hence, each node is an anchor.

Fig. 17.4 Dynamic
cluster-based tracking first
example

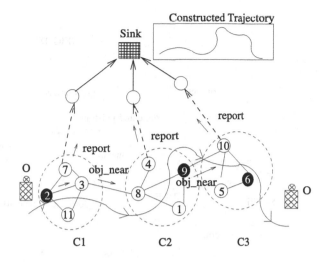

Fig. 17.5 Dynamic
cluster-based tracking second
example

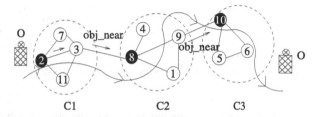

- Only nodes that have detected the object are included in the cluster and therefore participate in finding the coordinates of the object. In other words, any node that hears detection of an object by a neighbor node is not included in the cluster.
- CH is elected as the node with the minimum value of $\langle id, distance \rangle$, so that ties are broken by identifiers.
- The cluster that is formed is one-hop.

The algorithm proceeds in two phases. The first phase is the *CH Election Phase*, where until the duration *timeout1*, any node that has detected an object estimates its distance to the object and broadcasts its coordinates and the distance to the neighbors by the *detected* message. The node with the lowest $\langle id, distance \rangle$ is elected as the CH at the end of this phase, which broadcasts its identifier by the *clusterhead* message. The CH sends the coordinates of the object to the sink over the spanning tree T or a backbone structure that may be present, by the *report* message. It also sends the speed and direction to the first node in the estimated direction, which is forwarded to other nodes until it reaches the first node that is not a member of the cluster. The FSM of the proposed algorithm is shown in Fig. 17.6.

Each node initially is at IDLE state, and when a node detects an object O, it enters OBJ_FND state, estimates its distance to O, and broadcasts its identifier, its

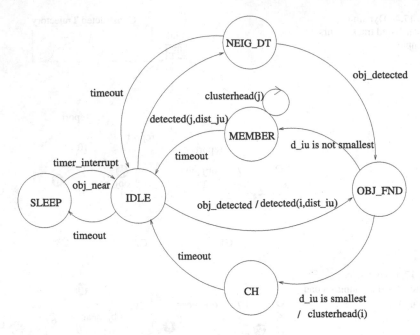

Fig. 17.6 Dynamic cluster-based tracking algorithm FSM

coordinates, and its estimated distance to the object by the *detected*($i, x_i, y_i, dist_{iO}$)
message. When a node that has not detected the object hears a neighbor by the
detected message, it enters the NEIGH_DT state with the expectation it may detect the object. If this does not happen, it goes back to the IDLE state after a
timeout. If a node detects an object at the NEIGH_DT state, it proceeds to the
OBJ_FND state, where nodes now decide on the CH of the cluster formed. If a
node finds after a timeout that it has the smallest ⟨*id, distance*⟩ pair among all the
nodes in OBJ_FND state, it declares itself as the CH by the *clusterhead* message.
Algorithm 17.2 (*Dclus_TT*) shows a sketch of the implementation of the above
scheme.

Further enhancements are possible such as providing a second phase of detection by the member nodes. In this case, any node that has detected the object in the first phase and the CH wait for another duration *timeout2* to further record any more detection of the object distance, which can then be sent
to CH. This second possible detection allows the speed and direction of the object to be computed in better accuracy, and at the end of this phase, the CH
may have the approximate coordinates, speed, and direction, which it sends to
the sink by the *report* message. The CH may aggregate data and send more re-
fined information such as a subtrajectory if more than two points become available. The message complexity of this algorithm is $O(n)$ as each detecting node
will send a constant number of *detected*, *clusterhead*, *report*, and *obj_near* messages.

Algorithm 17.2 *Dclus_TT*

 1: **states** *SLEEP, IDLE, NEIG_DT, OBJ_FND, MEMBER, CH, currstate*
 2: **tuples** *dists*[] ← ∅
 3: **message types** *detected, clusterhead, obj_near, report*
 4: **int** *count* ← 0; *my_CH*
 5:
 6: **loop**
 7:     **set** *timer* **to** *tval*
 8:     *currstate* ← *SLEEP*
 9:     *timer interrupt* or *obj_near* message
10:     *currstate* ← *IDLE*
11:     **while** ¬*timeout*1 **do**   ▷ clusterhead election phase
12:         **wait** for event   ▷ either node or a neighbor detects an object
13:         **case** *event_type* **of**
14:             *object_detected*:
15:                 **estimate** d_{iu}
16:                 *dists*[*count*] ← (x_i, y_i, d_{iu})
17:                 *count* ← *count* + 1
18:                 **broadcast** *detected*(x_i, y_i, d_{iu})
19:                 **if** *currstate* ≠ *OBJ_FND* **then** *currstate* ← *OBJ_FND*
20:             *detected*(x_j, y_j, d_{ju}) received:
21:                 **if** *currstate* = *IDLE* **then** *currstate* ← *NEIGH_DT*
22:                 *dists*[*count*] ← (x_j, y_j, d_{ju})
23:     **end while**
24:     **if** *count* ≥ 3 **then**
25:         **if** $i = \min\{j|dists[j, xj, y_j]\}$ **then**
26:             *currstate* ← *CH*
27:             *my_CH* ← *i*
28:             **broadcast** *clusterhead*
29:         **else if** *currstate* = *OBJ_FND* **then**
30:             **receive** *clusterhead*(*j*)
31:             *currstate* ← *MEMBER*
32:             *my_CH* ← *j*
33:         **end if**
34:     **end if**
35:     **if** *currstate* = *CH* **then**
36:         **compute** x_u, y_u from *dists* using Eq. (17.8)
37:         **predict** *speed, dir* from *dists*   ▷ estimate speed and direction of object
38:         **find** node *v* in direction *dir*
39:         **broadcast** *obj_near*(*speed, dir*)   ▷ warn neighbors of the coming object
40:         **send** *report*$(x_u, y_u, speed, dir)$ to *parent*   ▷ send info to sink via backbone
41:     **end if**
42: **end loop**

Fig. 17.7 Dynamic
tree-based tracking first
example

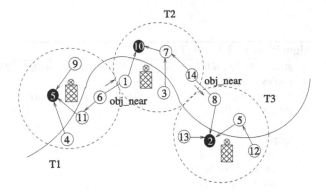

17.2.2 Tree-Based Approaches

Tree-based methods build a small tree in the vicinity of the moving object to estimate its coordinates. The nodes that detect the object elect a leader using an election
algorithms as in dynamic cluster-based schemes, the leader node becomes the *root*
of the tree, and the root forms a tree with few hops around the target. The estimated
distances of the object are forwarded to the root, which then uses trilateration to
estimate the coordinates of the object. If the root knows the previous direction and
speed of the object, it can estimate its possible next direction and warn the first
node in the estimated direction, which may then become the root of the next tree
to be formed. Figure 17.7 displays this method where an object O is detected by
the nodes 5, 4, 11, and 6. Node 5 has the shortest distance to O and is therefore
elected as the root of the tree T_1. The remaining nodes 4, 9, and 11 are the children
of the root, and node 6 is two hops away. Node 5 calculates the coordinates, speed,
and direction of O and sends this to the sink along the overlay tree T and also to
its children. Node 6 finds that it has a neighbor 1 in the direction of the object and
informs the coming object by the *obj_near* message, and the second tree T_2 along
the trajectory is formed.

   Alternatively, the nonmember node that is awaken by the *obj_near* message may
be assigned as the root of the next tree along the trajectory of the object as shown in
Fig. 17.8.

   The algorithm we propose to accomplish tree-based tracking is similar to
Dclus_TT and is called *Dtree_TT*. It also has two phases as before, and the first
phase is concluded by the election of the *root*. However, we assumed that the tree
formed around the object has a maximum depth of 2 to allow that any nonneighbor
nodes that have detected the object to cooperate to localize its coordinates. For this
reason, the *detected* message now has the extra *hops* field, which is set to 2 by any
node sending *detected* for the first time and is decremented by any node that hears
it and is broadcast until *hops* becomes 0. At the end of the first phase, *root* forms
a tree around it by the *probe* message similar to the *Flood_ST* algorithm we saw
in Sect. 4.2. The FSM of *Dclus_TT* is shown in Fig. 17.9, where a node that has
detected an object is assigned to one of the ROOT, INTERM, or LEAF states at the

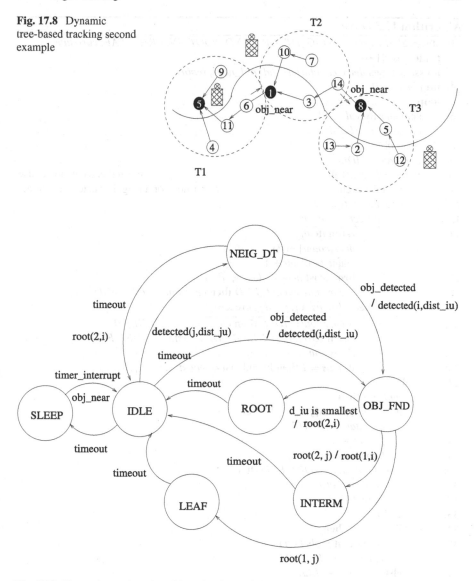

Fig. 17.8 Dynamic tree-based tracking second example

Fig. 17.9 Dynamic tree-based tracking algorithm FSM

end of first phase. After *timeout2* duration, all nodes go back to the SLEEP state as before.

Tree-based tracking this way provides nonadjacent nodes to cooperate to track an object in a two-hop structure. Algorithm 17.3 (*Dtree_TT*) shows one way of achieving tree-based tracking assuming that all nodes know their coordinates and their one-hop neighbor coordinates. The message complexity of this algorithm is $O(n)$ as each node transmits a constant number of messages.

Algorithm 17.3 *Dtree_TT*

1: **states** *SLEEP, IDLE, NEIG_DT, OBJ_FND, ROOT, INTERM, LEAF, currstate*
2: **tuples** *dists*[] ← ∅
3: **message types** *detected, clusterhead, obj_near, report*
4: **int** *count* ← 0
5: **loop**
6:    set *timer* to *tval*
7:    *currstate* ← *SLEEP*
8:    *timer interrupt*
9:    *currstate* ← *IDLE*
10:   **while** ¬*timeout*1 **do**   ▷ clusterhead election phase
11:      **wait** for event   ▷ either node or a neighbor detects an object
12:      **case** *event_type* **of**
13:         *object_detected*:
14:            estimate d_{iu}
15:            *dists*[*count*] ← (x_i, y_i, d_{iu})
16:            *count* ← *count* + 1
17:            **broadcast** *detected*(x_i, y_i, d_{iu})
18:            **if** *currstate* ≠ *OBJ_FND* **then** *currstate* ← *OBJ_FND*
19:         *detected*$(hops, x_j, y_j, d_{ju})$ received:
20:            **if** *currstate* = *IDLE* **then** *currstate* ← *NEIGH_DT*
21:            **else if** *currstate* = *NEIGH_DT* *currstate* ← *OBJ_FND*
22:            *dists*[*count*] ← (x_j, y_j, d_{ju})
23:            **if** *hops* = 2 **then broadcast** *detected*$(1, x_j, y_j, d_{ju})$
24:   **end while**
25:   **if** *count* ≥ 3 **then**
26:      **if** $i = \min\{j | dists[j, xj, y_j]\}$ **then**
27:         *currstate* ← *ROOT*
28:         *parent* ←⊥
29:         **broadcast** *root*(2, *i*)
30:      **else if** *currstate* = *OBJ_FND* **then**
31:         **receive** *root*(*hops*, *i*)
32:         *parent* ← *j*
33:      **end if**
34:      **if** *hops* = 2 **then**
35:         **broadcast** *probe*(1, *i*)
36:         *currstate* ← *INTERM*
37:      **else** *currstate* ← *LEAF*
38:         *parent* ← *j*
39:      **end if**
40:   **end if**
41:   **if** *currstate* = *ROOT* **then**
42:      **compute** x_u, y_u from *dists* using Eq. (17.8)
43:      **predict** *speed, dir* from *dists*   ▷ estimate speed and direction of object
44:      **find** node *v* in direction *dir*
45:      **broadcast** *obj_near*(*speed, dir*)   ▷ warn neighbors of the coming object
46:      **send** *report*$(x_u, y_u, speed, dir)$ to *parent*   ▷ send info to sink via backbone
47:   **end if**
48: **end loop** *forever*

17.2.3 Prediction-Based Approaches

In order to predict the next location of an object, *linear prediction*, which assumes linear motion and estimates the next location from the previous locations may be used. The three important methods for prediction are Kalman Filter, Extended Kalman Filter, and Particle Filtering methods.

Kalman Filter (KF) is a method to estimate the state of a linear system at a given time from the estimated previous state and current measurements [7]. It uses the dynamic model of the system, the control inputs, and multiple measurements to provide an estimate of the state of the system. It is mainly a sensor and data fusion algorithm, which has been extensively used in navigation and guidance of vehicles.

The *Extended Kalman Filter (EKF)* is the nonlinear version of KF, and it linearizes the original nonlinear filter dynamics around the previous state estimates [8]. In EKF, the state transition and observation models may not be linear functions of the state. The EKF is considered as the standard method in nonlinear state estimation. For tracking targets with nonlinear motion in a WSN, EKF has limitations, and information exchange in this environment may be complicated. However, sequential Monte Carlo methods, called particle filtering (PF), that estimate nonlinear and/or non-Gaussian dynamic processes may be a favorable choice [6].

In PF, target tracking is considered as a dynamic state estimation problem, and an approximation to the optimal solution is searched. In this method, the posterior probability density function (pdf) of Bayesian estimation is represented with discrete samples (or particles) with associated weights. The algorithm performs the following at each iteration. First, particles from a proposal distribution are drawn and assigned the corresponding normalized weights; a resample may also be formed, and finally the estimate is formed as the weighted average of the particles. Particle filters require significant computation, and hence distributed algorithms to perform particle filtering may be challenging.

17.2.4 Lookahead Target Tracking

Most of the tracking algorithms predict the location of the target for the next short duration of time. If object location is determined at time t, the contemporary algorithms estimate its location at $t + 1$ and send warning messages to the nodes in the computed direction so that they are awaken and possibly organized in anticipation of the coming object. However, if the object moves with high speed, the awakening of the nodes in the next region may not provide the required accuracy for the trajectory estimation.

Alaybeyoglu et al. [1] proposed a dynamic cluster-based tracking algorithm, called *lookahead target tracking*, in which k future locations of the target are predicted, and k nodes, possible future clusterheads, are awaken. The value of k is dependent on the speed of the object and increases with high speeds. Figure 17.10 displays the general idea behind lookahead tracking, where a series of trees T_1 at

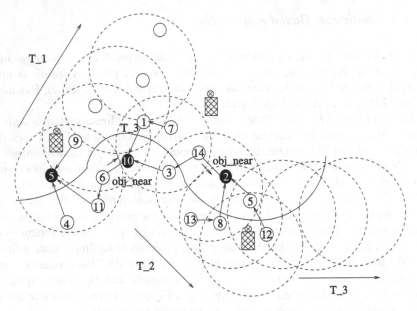

Fig. 17.10 Lookahead target tracking example

point A, T_2 at point B, and T_3 at point C are formed with the estimation of the target's future location with the k value of 4. Alaybeyoglu et al. achieved to track objects at speeds of up to 100 m/s with the described lookahead method combined with the particle filtering prediction method.

17.3 Chapter Notes

Target tracking is a fundamental WSN application, and there are considerable research efforts about various aspects of tracking. In general, three main approaches for awakening the nodes along the target trajectory are as follows. First, only one sensor nearest to the predicted destination can be awaken. Second, all nodes on the route of the target to the destination may be awaken, and third, all nodes around the route to the destination are awaken [9, 10]. The third method clearly provides better estimation but has high energy requirements due to the wakening of many redundant nodes. We have seen algorithms in this chapter that estimate routes based on the modified third method, where only a subset of nodes on the route to the destination are awaken.

An example of a tree-based scheme is the Dynamic Convoy Tree-based Collaboration (DCTC) [16], where sensor nodes form a tree around the object, and the dynamic tree is modified by adding or deleting nodes based on the trajectory of the object. The object route is gathered by the tree nodes and sent to the sink. A dynamic cluster-based algorithm is proposed in [5], where the network consists of high-capability sensors and normal nodes. The high-capability sensors act as the CHs, and when an object is detected, gather information from the normal sensors to send to the sink. Prediction-based methods have been used in [14, 15, 17], assuming that the speed and direction of the moving object do not change in the next few seconds. Alaybeyoglu et al. provided a cone-based dynamic clustering method where nodes that are in a cone-shaped region along the estimated route of a target are awaken. They showed that this method, combined with particle filter estimation, provides target estimates with high accuracy for highly dynamic targets having nonlinear motion [2].

Target tracking will probably remain as an active area of research in sensor network applications for many years as it has numerous diverse applications from habitat monitoring to intruder detection to military applications. Our general conclusion is that tracking objects require complicated systems that should have several components, some of which are described in this chapter as localization, topology constructs, and prediction techniques. Another important related area of research is multi-target tracking, where more than one target, their movement pattern, and rendezvous points are tracked and predicted.

17.3.1 Exercises

1. Compare RSSI, TDoA, and AoA methods in terms of precision and implementation costs involved.
2. The algorithm *Dclus_TT* is to be enhanced so that a second phase of detection is provided as described in Sect. 18.2. Provide a pseudocode for this modification.
3. Provide a pseudocode for the same modification to algorithm *Dtree_TT* as in Exercise 2.
4. In dynamic tree or cluster algorithms, the next CH or tree root can be elected as the next node in the current moving direction of the target or the closest node to the target that has detected it, in that region. Compare these two methods discussing their relative advantages and disadvantages.
5. Discuss briefly the effect of the velocity of the target to be tracked on the trajectory estimated using lookahead tracking.

References

1. Alaybeyoglu A, Erciyes K, Dagdeviren O, Kantarci A (2010) A dynamic distributed algorithm for tracking fast moving targets in wireless sensor networks. IETE Tech Rev 27(1):46–53
2. Alaybeyoglu A, Erciyes K, Kantarci A (2013) An adaptive cone based distributed tracking algorithm for a highly dynamic target in wireless sensor networks. Int J Ad Hoc Ubiq Comput 12(5):98–119
3. Bachrach J, Taylor C (2005) Localization in sensor networks. In: Stojmenovic I (ed) Handbook of sensor networks: algorithms and architectures. Wiley, New York, Chapter 9

4. Bulusu N, Heidemann J, Estrin D (2000) GPS-less low cost outdoor localization for very small devices. IEEE Pers Commun 7:28–34
5. Chen WP, Hou J, Sha L (2003) Dynamic clustering for acoustic target tracking in wireless sensor networks. IEEE Trans Mob Comput 3(3):258–271
6. Doucet A, de Freitas N, Gordon N (eds) (2001) Sequential Monte Carlo methods in practice. Statistics for engineering and information science. Springer, New York
7. Gelb A (1974) Applied optimal estimation. MIT Press, Cambridge
8. Grewal MS, Andrews P (1993) Kalman filtering: theory and practice. Prentice Hall, New York
9. Gui C, Mohapatra P (2004) Power conservation and quality of surveillance in target tracking sensor networks. In: Proc 10th annual international conference on mobile computing and networking, MobiCom '04, pp 129–143
10. Gupta R, Das SR (2003) Tracking moving targets in a smart sensor network. In: Vehicular technology conference, IEEE VTC 2003, vol 5, pp 3035–3039
11. Kim SY, Kwon OH (2005) Location estimation based on edge weights in wireless sensor networks. J Korea Inform Commun Soc 30
12. Niculescu D, Nath B (2001) Ad hoc positioning system (APS). In: Proc IEEE GLOBECOM 2001, pp 2926–2931
13. Whitehouse K, Culler D (2003) Macro-calibration in sensor/actuator networks. Mob Netw Appl 8(4):463–472
14. Xu Y, Winter J, Lee W-C (2004) Dual prediction-based reporting for object tracking sensor networks. In: Proc mobile and ubiquitous systems: networking and services, MOBIQUITOUS 2004, pp 154–163
15. Yingqi X, Lee W-C (2004) Prediction-based strategies for energy saving in object tracking sensor networks. In: Proc fifth IEEE international conference on mobile data management (MDM'04)
16. Zhang W, Cao G (2004) DCTC: dynamic convoy tree-based collaboration for target tracking in sensor networks. IEEE Trans Wirel Commun 3(5):1689–1701
17. Zhao F, Shin J, Reich J (2002) Information-driven dynamic sensor collaboration for tracking applications. IEEE Signal Process Mag 19(2):61–72

Chapter 18
ASSIST: A Simulator to Develop Distributed Algorithms

Abstract We describe a simple simulator, called *ASSIST*, based on POSIX threads to develop distributed algorithms in this chapter. ASSIST is easy to learn and implement and can be used to test and verify distributed algorithms.

18.1 Introduction

Simulators are widely used to test, verify, and analyze the performances of distributed algorithms. We have briefly described simulators that are in use for MANETs and WSNs such as *ns*2, OMNET++, and TOSSIM in Chap. 14. In this chapter, we show the implementation of a simple simulator called ASSIST (A Simple Simulator based on Threads) that we have designed and used in teaching of distributed algorithm courses. ASSIST is a simple software package written in C and based on POSIX threads. Two important functions of ASSIST are the memory management and interprocess communication. Memory management is handled by buffer pools, which are statically allocated at system initialization. Memory management in this way prevents overflows and provides an efficient method to manage memory space.

ASSIST provides the interprocess communication module for the nodes of the distributed system to communicate with each other. Each node of the distributed system is simulated by a POSIX thread, which communicates with its neighbors using the interprocess communication procedures to simulate communications over the network. As the distributed algorithms are symmetric in most of the cases, providing the distributed algorithm code for a single thread is adequate. For example, for the spanning tree construction using flooding (*Flood_ST*), we can write a single thread and invoke this thread as many times as the number of nodes in the network. However, we need to define the neighborhood of the nodes so that each node is allowed to communicate with its neighbors only. The memory management module and a simple interprocess communication using semaphores and mutex variables are described below. The actual C code for ASSIST is given in Appendix B.

18.2 Memory Management by Buffer Pools

Memory management in ASSIST is provided by *buffer pools*. A buffer pool is a data structure that contains the physical space for required items of data units called

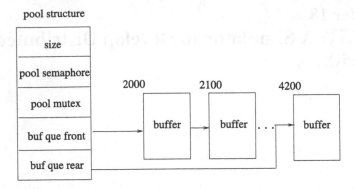

Fig. 18.1 Buffer pool data structure

buffers. This data structure has an array of buffers managed by the *pool* structure. This structure has a semaphore called *poolsem*, and a thread that needs a free buffer has to perform a *wait* on this semaphore to check the availability of a free buffer. The head of the queue is stored in *bufque_front* and the address of the last free buffer is kept at *bufque_rear*. Mutual exclusion to shared *pool* structure is provided by the lock variable *pool mutex*. There are a total of *size* number of buffers allocated, and each buffer initially is linked to the next one as shown in Fig. 18.1. The pool semaphore *poolsem* is initialized to the number of empty buffers. Obtaining a free buffer from the pool is performed in a first-in-first-out manner.

Two procedures required for buffer management are as follows:

- *get_buf* (*pool type pointer pp*): Calling thread waits on the pool semaphore, and it retrieves the buffer address at the top of the pool when signalled.
- *put_buf* (*pool type pointer pp*, *buffer pointer bp*): Calling thread puts the address of the buffer to the rear of the queue and signals the pool semaphore to activate a waiting process.

Typically in a single-processor system, a producer process will invoke *get_buf* to obtain a free buffer, fill this buffer with data, and send it to the consumer process. The consumer process will receive the buffer, use the data in this buffer, and return the buffer to the pool by the *put_buf* procedure. The communication between the producer and the consumer processes can be handled asynchronously in this manner. The buffer management module of ASSIST provides the initialization procedure *init_pool*, which initializes the pool and its semaphores. Also, the *check_pool* procedure is used by a process that wants to check the pool before obtaining a buffer from the pool to prevent waiting on the semaphore.

18.3 Interprocess Communication

Interprocess communication in ASSIST is provided by the data structure *fifo*, which simulates the *first-in-first-out* structure of the communication channels. Each *fifo*

fifo structure

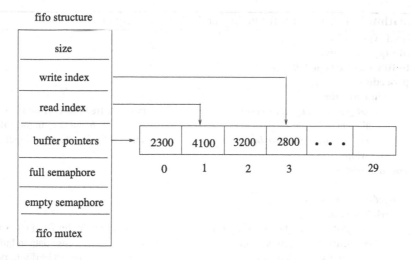

Fig. 18.2 Fifo data structure

has space allocated to hold certain number of buffer pointers up to the *size* of the *fifo*. The *empty semaphore*, which shows the number of empty buffer pointer locations, is initialized to the size of fifo, and the *full semaphore*, which shows the current unread buffer pointers, is initialized to 0. The two indexes, the *read index* and the *write index*, show the next location in the buffer pointer array to read and write, respectively, as shown in Fig. 18.2. The mutual exclusion variable *fifo mutex* provides a single access to the *fifo* structure at any time. For this example *fifo*, 30 buffer address spaces are allocated, the read index is 1, meaning that the next buffer address that will be read is 4100, and the write index is 3, showing that the next write will be to the third array location, overwriting data at address 2800. The full semaphore has the value 2, showing that there are two locations that can be read (1 and 2) before being blocked on this semaphore, and the empty semaphore, which shows the empty locations, has the value 28, meaning that there may be 28 (locations 3 to 29 and 0) consecutive write operations to this fifo before being blocked on this semaphore.

The procedures to send a message to and receive a message from a fifo are as follows:

- *write_fifo(fifo pointer fp, buffer pointer bp)*: If *fifo* is full, the calling process is blocked on the full semaphore of the *fifo*. Otherwise, the address of the buffer is written to the next available place in the *fifo* shown by the write index.
- *read_fifo(fifo pointer pp)*: It returns the next available buffer address from the *fifo* shown by the read index. If there are no messages in the *fifo* as shown by the 0 value of the empty semaphore, the caller is blocked.

The initialization procedure *init_fifo* initializes the *fifo* and its semaphores. Algorithm 18.1 shows a simple producer consumer example where a producer process obtains a buffer from the pool, fills it with data, and sends it to the input *fifo*

Algorithm 18.1 Procedures for finding and allocating neighbors

---

 1: **pool type** *pool*
 2: **fifo type** *cons_fifo*
 3: **buffer pointer type** *buf_ptr*
 4: **procedure** *Producer*
 5:     **while** *true* **do**
 6:         *buf_ptr* ← *get_buf* (&*pool*)   ▷ obtain a free buffer from the pool
 7:         **fill** data to buffer   ▷ put data in the buffer
 8:         *write_fifo*(&*cons_fifo*, *buf_ptr*)   ▷ place buffer in consumer input *fifo*
 9:     **end while**
10: **end procedure**
11:
12: **procedure** *Consumer*
13:     **while** *true* **do**
14:         *buf_ptr* ← *read_fifo*(&*cons_fifo*)   ▷ get the buffer from input *fifo*
15:         **consume** data from *buf_ptr*   ▷ use data in buffer
16:         *put_buf* (&*pool*, *buf_ptr*)   ▷ return buffer to pool
17:     **end while**
18: **end procedure**

(*cons_fifo*) of the consumer. The consumer receives the buffer from its *fifo* by the *read_fifo* procedure, uses the data in the buffer, and returns it to the pool.

18.4 Sliding-Window Protocol Implementation

In this section, we will illustrate the use of *ASSIST* by a simple sliding-window data link protocol. This protocol called *Go-back-N* is used for flow control and error control between a sending node and a receiving node in computer networks. It is typically used in data link but can also be used in other layers such as the transport layer. The sending node in this protocol may send a certain number of frames of data from a window of frames, without getting an acknowledgement from the receiver. When all the frames in the current window are sent, the sender is blocked, waiting for an acknowledgement from the receiver. The receiver may acknowledge a number of frames by a single acknowledgement (*ACK*). When an acknowledgement is received, the sender shifts its window as many as the number of frames acknowledged. In *Go-back-N* protocol, if there is an error in a sent frame, the receiver sends a negative acknowledgement (*NACK*) for this frame, and all of the sent frames starting from the erroneous frame are retransmitted. This type of flow control is necessary to enable the synchronization between a sender and a receiver to prevent buffer overflows. Figure 18.3 shows an example operation of the protocol.

ASSIST may be used to realize this protocol, where the sending thread sends the contents of an input file to the receiving thread. The sending thread fills the array *swindow* of size *N* with buffer addresses obtained from the buffer pool, where the indexes of buffer pointers in *swindow* are the sequence numbers of the frames to

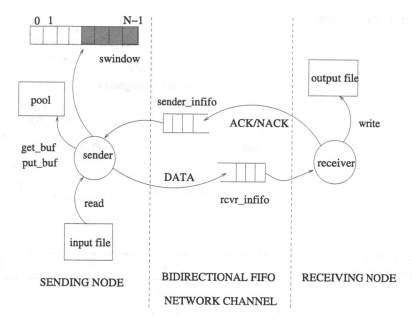

Fig. 18.3 Go-back-N implementation with ASSIST

be transmitted. The sender gets an address of a buffer from the available window, reads 10 bytes of data from the input file, writes this data to the buffer, and writes the buffer address to the receiver fifo. Whenever an ACK is received, more buffer addresses are obtained from the pool to fill the window as shown in Fig. 18.3. The actual C code for the protocol is given in Sect. B.1.

18.5 Spanning Tree Construction

In this section, we will describe the implementation of the spanning tree construction algorithm with termination detection, *Term_ST*, of Sect. 4.3. The FSM of this algorithm is presented here in Fig. 18.4 with the addition of actions to be invoked. This algorithm starts by a single initiator, called the *root*, which sends *probe* messages to all of its neighbors. Any node that receives a probe for the first time marks the sender as its parent, enters *XPOLRD* state, and sends the probe message to all its neighbors except the parent. When a node receives *ack* or *reject* messages from all its neighbors except the parent, it sends an *ack* message to its parent. In this way, all *ack* messages are convergecast to the root, which marks the end of the algorithm.

In order to implement *Term_ST* using *ASSIST*, we would first need to provide the state table for the FSM of the algorithm shown in Fig. 18.4. The state table as shown in Table 18.1 has the states *IDLE* and *XPLORD* as rows and the inputs *probe*, *ack*, and *reject* as the columns. When an input is received, the related action is decided according to the entry of the table specified by the current state and input. For

Fig. 18.4 *Term_ST FSM*

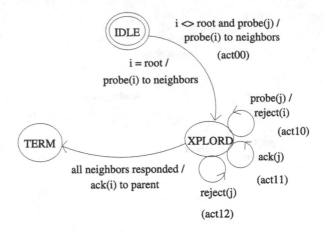

Table 18.1 State table for the parity checker

	probe	ack	reject
IDLE	act00	–	–
EVEN	EVEN	ODD	–
XPLORD	act10	act11	act12

example, receiving *probe* input at *XPLORD* state would activate *action*10, which would send a *reject* message to the sender as the node has already received a *probe* message.

18.5.1 Data Structures and Initialization

The following constants and data structures were defined for FSM states (IDLE, XPLORD), the message types (*probe*, *ack*, *reject*), and the FSM type, which is a two-dimensional array holding a function pointer that is initialized with the corresponding action entry. Current states, neighbors, and children for each node are declared as global arrays where the ith entry in these arrays correspond to the values for the ith process. This way, actions called from any process can modify them using the identity of the process. Also, each node has an input message queue specified by its identifier-based entry in the array *fifos*.

```
#define PROBE     0    // message types
#define ACK       1
#define REJECT    2
#define IDLE      0    // states
#define XPLORD    1
#define TERM      2
#define N         6    // number of nodes
#define ROOT      3    // root node
```

```
int parents[N]={0}, currstates[N]={0};
int neighs[N][N+3]={...}, others[N][N+3]={0},
childs[N][N+3]={0};

typedef int   (*fnptr_t)();
typedef fnptr_t fsm_t[2][2];      // FSM type declaration
fifo_t fifos[N];                  // fifos
pool_t mypool;                    // buffer pool declaration
```

18.5.2 The Algorithm Thread

The actual thread that is activated as the number of nodes N in the distributed system is shown below. It starts by receiving its identifier through the thread creation function executed by the main program. The reception of the identifier is important as it forms the index to manipulate the related entry in the global data for a particular thread. For example, neighbors of thread k are stored in the global data *neighs*[k][$N + 1$]. The user should initialize the *neighs* structure to reflect the network topology, or the thread itself can do the initialization when invoked.

The designated *root* thread starts by sending the *probe* message to all its neighbors. All the remaining threads start by waiting to receive a message from their *fifos*. The main body of the FSM is simply to activate action determined by the current state as the row and the received message type as the column of the state table. The main loop is executed indefinitely until the current state of the thread is *TERM*, which signals the end of the execution as shown below.

```
void CRobs_node(int *mp)
{  fsm_t my_fsm;          // declare my FSM
   int currstate = IDLE;
   initialize()
   int me=*mp;bufptr bp;


   if (me==ROOT)          // if i am root, start the algorithm
      while(neighs[j]!=-1) // for all neighbors
         { bp = get_buf(pool);   // get a buffer
           bp-> type = PROBE;    // set type as PROBE
           bp->sender = me;      // insert my id
           write_fifo(fifos[me], bp);} // send it to neighbor

   while (currstates[me]!=TERM)   // loop until state = TERM
   {
     bp=read_fifo(&fifos[me]);
     (*myfsm[currstates[me]][bp->type])(me,bp);
   }
}
```

Fig. 18.5 A spanning tree
formed by *Term_ST*
algorithm

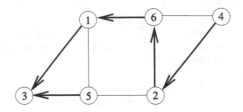

Figure 18.5 displays a possible spanning tree formed by the algorithm in a network of six nodes with identities $1, \ldots, 6$, where the root node is 3. As threads are activated randomly by most operating systems, they may execute in any order, resulting in possibly different spanning trees at each run. For this example, it may be seen that node 5 is scheduled later than others by the operating system, and hence sending of the *probe* message by the node 5 is delayed causing it to reach node 2 later than the *probe* message sent by node 6 resulting in node 6 becoming parent of node 2. The full implementation code in C is shown in Sect. C.2

18.6 Chapter Notes

We showed a simple simulator to run distributed algorithms. Most of the contemporary simulators require a learning and experimenting time, which is not trivial. For courses on distributed algorithms and ad hoc networks, this allocation of learning time is almost impossible, which makes it difficult for the instructors to teach the students the implementation of distributed algorithms.

ASSIST grew out of the necessity for such a condition. It is simple so that it can be learned and experienced within hours in contrast to months of many contemporary simulators. It has been successfully used in senior/graduate level courses on distributed systems and algorithms at various universities in the world, to implement programming assignments and projects as the one described in Sect. 18.5.

Our general conclusion is that ASSIST can aid designing, development, testing, and verification of distributed algorithms for ad hoc networks and general computer networks. One major drawback with ASSIST is that as the number of nodes in the simulated network increase, configuration of the neighbors of a node will not be easy. For example, to simulate a WSN with n nodes ($n \gg 1$), the establishment of neighbors for n nodes would require substantial lines of initialization code to be added to assign neighbors to nodes. A graphical user interface may be provided in future to ease such a problem. ASSIST was primarily used for designing, testing, and verification. For performance evaluation such as measuring durations of executions for a large network, the widely used network simulator *ns2* would probably be a better choice. However, in measuring relative performances of few distributed algorithms on the same testbed with the same network using ASSIST would provide clues about the relative favorableness of these algorithms.

Fig. 18.6 Example graph for Project 1

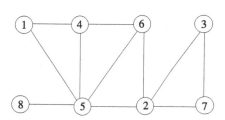

Fig. 18.7 Example graph for Project 2

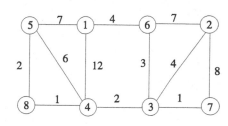

Fig. 18.8 Example graph for Project 3

18.6.1 Projects

1. It is required that each node in a network should be informed the identifiers of all nodes in this network. Assuming there is already a constructed spanning tree as in Fig. 18.6, provide a distributed algorithm where the identifiers of children of all nodes in this tree are convergecast to the root node 3, which in turn broadcasts these identifiers to all nodes. Write this algorithm using ASSIST by first initializing the *neighs* data structure for this network.
2. Implement the asynchronous routing algorithm called *Chandy_APSP* of Sect. 7.6 using ASSIST based on an FSM. Find the resulting routes from node 5 in the example network of Fig. 18.7.
3. Implement the greedy distributed vertex coloring algorithm called *Rank_Vcol* of Sect. 9.3 using ASSIST based on an FSM and find the resulting colors in the example network of Fig. 18.8.

Appendix A
Pseudocode Conventions

A.1 Introduction

In this part, the pseudocode conventions for writing an algorithm are presented. The conventions we use follow the modern programming guidelines and are similar to the ones used in [1] and [2]. Every algorithm has a name specified in its heading, and each line of an algorithm is numbered to provide citation. The first part of an algorithm usually starts by its inputs. Blocks within the algorithm are shown by indentations. The pseudocode conventions adopted are described as data structures, control structures, and distributed algorithm structure as follows.

A.2 Data Structures

Expressions are built using constants, variables, and operators as in any functional programming language and yield a definite value. *Statements* are composed of expressions and are the main unit of executions. All statements are in the form of numbered lines. Declaring a variable is done as in languages like Pascal and C, where its type precedes its label with possible initialization as follows:

$$\textbf{set of int} \quad \textit{neighbors} \leftarrow \varnothing$$

Here we declare a set called *neighbors* of a vertex in a graph each element of which is an integer. This set is initialized to $\varnothing$ (empty) value. The other commonly used variable types in the algorithms are *boolean* for boolean variables and *message types* for the possible types of messages. For assignment, we use $\leftarrow$ operator, which shows that the value on the right is assigned to the variable in the left. For example, the statement

$$a \leftarrow a + 1$$

increments the value of the integer variable a. Two or more statements in a line are separated by semicolons and comments are shown by $\triangleright$ symbol at the end of the

K. Erciyes, *Distributed Graph Algorithms for Computer Networks*,
Computer Communications and Networks, DOI 10.1007/978-1-4471-5173-9,
© Springer-Verlag London 2013

Table A.1 General
algorithm conventions

Notation	Meaning
$x \leftarrow y$	assignment
$=$	comparison of equality
$\neq$	comparison of inequality
true, *false*	logical true and false
null	nonexistence
$\triangleright$	comment

Table A.2 Arithmetic and
logical operators

Notation	Meaning
$\neg$	logical negation
$\wedge$	logical and
$\vee$	logical or
$\oplus$	logical exclusive-or
x/y	x divided by y
$x \cdot y$ or xy	multiplication

line as follows:

$$1 : a \leftarrow 1; c \leftarrow a + 2; \qquad \triangleright \quad c \text{ is now } 3$$

General algorithmic conventions are outlined in Table A.1.

Table A.2 summarizes the arithmetic and logical operators used in the text with their meanings.

Sets instead of arrays are frequently used to represent a collection of similar variables. Inclusion of an element u to a set S can be done as follows:

$$S \leftarrow S \cup \{u\}$$

and deletion of an element v from S is performed as follows:

$$S \leftarrow S \setminus \{v\}$$

Table A.3 shows the set operations used in the text with their meanings.

A.3 Control Structures

In the sequential operation, statements are executed consecutively. Branching to another statement can be done by *selection* described below.

Table A.3 Set operations

Notation	Meaning
$|S|$	cardinality of S
$\varnothing$	empty set
$u \in S$	u is a member of S
$S \cup R$	union of S and R
$S \cap R$	intersection of S and R
$S \setminus R$	set subtraction
$S \subset R$	S is a proper subset of R
$\max / \min S$	maximum/minimum value of the elements of S
$\max / \min\{\dots\}\ S$	maximum/minimum value of a collection of values

---

Algorithm A.1 if–then–else structure

```
 1: if condition then                                        ▷ first check
 2:     statement1
 3:     if condition2 then                               ▷ second (nested) if
 4:         statement2
 5:     end if                                           ▷ end of second if
 6: else if condition3 then                              ▷ else if of first if
 7:     statement3
 8: else
 9:     statement4
10: end if                                               ▷ end of first if
```

A.3.1 Selection

Selection is performed using conditional statements, which are implemented using *if–then–else* in the usual fashion, and indentation is used to specify the blocks as shown in the example code segment (see Algorithm A.1).

In order to select from a number of branches, *case–of* construct is used. The expression within this construct should return a value that is checked against a number of constant values, and the matching branch is taken as follows:

1. **case** *expression* **of**
2. $constant_1$: $statement_1$
3. $\vdots$
4. $constant_n$: $statement_n$
5. **end case**

A.3.2 Repetition

The main loops in accordance with the usual high-level language syntax are the *for*, *while*, and *loop* constructs. The *for–do* loop is used when the count of iterations can be evaluated before entering the loop as follows:

1. **for** $i \leftarrow 1$ **to** n **do**
2. $\vdots$
3. **end for**

The second form of this construct is the *for all* loop, which arbitrarily selects an element from the set specified and iterates until all members of the set are processed as shown below, where a set S with three elements and an empty set R are given, and each element of S is copied to R iteratively.

1. $S \leftarrow \{3, 1, 5\}; R \leftarrow \varnothing$
2. **for all** $u \in S$ **do**
3. $R \leftarrow R \cup \{u\}$
4. **end for**

For the indefinite cases where the loop may not be entered at all, the *while–do* construct may be used, where the boolean expression is evaluated, and the loop is entered if this value is true as follows:

1. **while** *boolean expression* **do**
2.   *statement*
3. **end while**

A.4 Distributed Algorithm Structure

Distributed algorithms have significantly different structures than the sequential algorithms as their execution pattern is determined by the type of messages they receive from their neighbors. For this reason, the general distributed algorithm pseudocode usually includes a similar structure to the algorithm template shown in Algorithm A.2.

In this algorithm structure, there may be n types of messages, and the type of action depends on the type of message received. For this example, the **while–do** loop executes as long as the value of the boolean variable *flag* evaluates to *true*. Typically, a message received by this node (i) at some point triggers an action that changes the value of the *flag* variable to *true*, which then results in this termination of the loop. In another frequently used distributed algorithm structure, the **while–do** loop is executed forever, and one or more of the actions should provide *exit* from this endless *while* loop as shown in Algorithm A.3.

The indefinite structure of this loop type makes it suitable for distributed algorithms where the type of message, in general, cannot be determined beforehand.

Algorithm A.2 Distributed Algorithm Structure 1

```
 1: int i, j                      ▷ i is this node; j is the sender of the current message
 2: while ¬flag do                            ▷ all nodes execute the same code
 3:     receive msg(j)
 4:     case msg(j).type of
 5:             type_1:   Action_1
 6:             ...   :   ...
 7:             type_n:   Action_n
 8:     if condition then
 9:         flag ← true
10:     end if
11: end while
```

Algorithm A.3 Distributed Algorithm Structure 2

```
1: while forever do
2:     receive msg(j)
3:     case msg(j).type of
4:             type_1:   Action_1: if condition₁ then exit
5:             ...   :   ...
6:             type_x:   Action_1: if condition_x then exit
7:             type_n:   Action_n
8: end while
```

References

1. Cormen TH, Leiserson CE, Rivest RL, Stein C (2001) Introduction to algorithms. MIT Press, Cambridge
2. Smed J, Hakonen H (2006) Algorithms and networking for computer games. Wiley, New York. ISBN 0-470-01812-7.

Appendix B
ASSIST Code

B.1 Buffer Pool Management

```
/*      assist.c  */
#include <stdio.h>
#include <fcntl.h>
#include <pthread.h>
#include <semaphore.h>
#include <stdlib.h>
#include <sys/types.h>


/****************************************************
          pool data structure
****************************************************/

#define       POOL_SIZE         20
                                /* size of allocated space */
#define       ERR_POOLEMPTY    -1
#define       ERR_POOLFULL     -2

typedef struct buffer *bufptr;
typedef struct buffer {
          int data;
          bufptr next;
          int sender;
          int type;
          } buf_t;

typedef struct pool *poolptr;
typedef struct pool{      int state;
       int pool_size;
       sem_t poolsem;
       pthread_mutex_t poolmut;
```

K. Erciyes, *Distributed Graph Algorithms for Computer Networks*,
Computer Communications and Networks, DOI 10.1007/978-1-4471-5173-9,
© Springer-Verlag London 2013

```
                bufptr front;
                bufptr rear;
                buf_t bufs[POOL_SIZE];
                        } pool_t;
```

```
/******************************************************
          initialize a buffer pool
******************************************************/
```

```
int     init_pool (poolptr pp, int pool_length)

{    int i ;
     pp->pool_size=pool_length;
     sema_init(&pp->poolsem,pp->pool_size,USYNC_THREAD,0);
     mutex_init(&pp->poolmut,USYNC_THREAD,0);
     pp->front=&(pp->bufs[0]);
     for (i=0; i<pp->pool_size - 1; i ++)
         pp->bufs[i].next = &(pp->bufs[i+1]);
         pp->rear=&(pp->bufs[pp->pool_size-1]);
}
```

```
/******************************************************
             get a buffer from a pool
******************************************************/
```

```
bufptr get_buf(poolptr pp)

{    bufptr bp;
     sema_wait(&pp->poolsem);
     mutex_lock(&pp->poolmut);
     bp=pp->front;
     pp->front=bp->next;
     mutex_unlock(&pp->poolmut);
     return(bp); }
```

```
/******************************************************
             put a buffer to a pool
******************************************************/
```

```
int put_buf(poolptr pp, bufptr bp)

{    bp->next=NULL;
     mutex_lock(&pp->poolmut);
     pp->rear->next=bp;
     pp->rear=bp;
     if (pp->front==NULL) pp->front=bp;
     mutex_unlock(&pp->poolmut);
     sema_post(&pp->poolsem); }
```

B.2 Interprocess Communication

```
/***************************************************
              fifo data structure
***************************************************/

#define        FIFO_SIZE            10
#define        N_FIFOS             10
#define        ALLOCATED            1
#define        ERR_FIFOEMPTY       -1
#define        ERR_FIFOFULL        -2


typedef struct fifo *fifoptr;
typedef struct fifo{    int state ;
            int fifo_size;
            int read_idx;
            int write_idx;
            sem_t  fullsem;
            sem_t emptysem;
            pthread_mutex_t fifomut;
            bufptr bufs[FIFO_SIZE];
            } fifo_t;

/***************************************************
              initialize a fifo
***************************************************/

int init_fifo(fifoptr fp)

{    int fifoid;
    fp->state=ALLOCATED;
    fp->fifo_size=FIFO_SIZE;
    sem_init(&fp->fullsem,0,0);
    sem_init(&fp->emptysem,fp->fifo_size,0);
    pthread_mutex_init(&fp->fifomut,0);
    fp->read_idx=0;
    fp->write_idx=0;
    return(fifoid); }

/***************************************************
              read a buffer from a fifo
***************************************************/

bufptr read_fifo(fifoptr fp)

{    bufptr bp;
    sem_wait(&fp->fullsem);
```

```
    pthread_mutex_lock(&fp->fifomut);
    bp=fp->bufs[fp->read_idx++];
    fp->read_idx %= fp->fifo_size;
    pthread_mutex_unlock(&fp->fifomut);
    sem_post(&fp->emptysem);
    return(bp); }


/***************************************************
              write a buffer to a fifo
***************************************************/

int write_fifo(fifoptr fp, bufptr bp)

{   sem_wait(&fp->emptysem);
    pthread_mutex_lock(&fp->fifomut);
    fp->bufs[fp->write_idx++]=bp;
    fp->write_idx %= fp->fifo_size;
    pthread_mutex_unlock(&fp->fifomut);
    sem_post(&fp->fullsem); }
```

Appendix C
Applications Using ASSIST

C.1 Sliding-Window Protocol Code

C.1.1 Data Structures and Initialization

```
#define DATA     0        // frame types
#define ACK      1
#define NACK     2
#define N        8        // sequence number range
#define W_SIZE N-1        // window size
#define N_DATA   10       // data block size

fifo_t sender_infifo, rcvr_infifo;  // communication channels
pool_t pool;                        // buffer pool
bufptr swindow[N];                  // sliding window

init_sys() { int i;                 // initialize system
          init_fifo(&sender_infifo);
          init_fifo(&rcvr_infifo);
          init_pool(&pool,20);
     }


void dl_sender()              // sending thread
{     FILE *fp1;
      bufptr bp;
      int count,len,wlength=N-1,last_frsent=-1,next_ackseq=-1;

      fp1=fopen("infile","r");
      while(1) {
          for(count=0;count<wlength;count++)
          {   bp=get_buf(&pool);
              last_frsent=(last_frsent+1)%N;
              swindow[last_frsent]=bp;
```

K. Erciyes, *Distributed Graph Algorithms for Computer Networks*,
Computer Communications and Networks, DOI 10.1007/978-1-4471-5173-9,
© Springer-Verlag London 2013

```
                len=fread(bp->data,sizeof(int),N_DATA,fp1);
                bp->seqnum=last_frsent;
                bp->length=len;
                calc_crc(bp);
                write_fifo(&rcvr_infifo,bp);
                if(len<N_DATA) exit;
            }
            bp=read_fifo(&sender_infifo);
            if (bp->seqnum > next_ackseq)
                wlength=bp->seqnum-next_ackseq;
            else wlength=N+bp->seqnum-next_ackseq;
        }
}


void dl_receiver()                 // receiving thread
{       FILE *fp2;
        int i,j,next_seq;
        bufptr bp, recvd[3];
        int count=0;

        fp2=fopen("outfile","w");
        while(1)
        {       for(i=0;i<3;i++)
                { recvd[i]=read_fifo(&rcvr_infifo);
                  fwrite(recvd[i]->data,sizeof(int),
                            recvd[i]->length,fp2);
                  if(recvd[i]->length<N_DATA) exit;
                }
                if(recvd[i]->length<N_DATA) exit;
                next_seq = (next_seq+1)%N;
                recvd[0]->seqnum=(recvd[2]->seqnum+1) % N;
                write_fifo(&rcvr_infifo,recvd[0]);
        }

}


main(){
        pthread_t sender_id, receiver_id;

        init_sys();
        pthread_create(&sender_id,NULL,(void*)dl_sender,
                        NULL);
        pthread_create(&receiver_id,NULL,(void*)dl_receiver,
                        NULL);

        pthread_join(sender_id,NULL);
        pthread_join(receiver_id,NULL);
}
```

Finally, we would need to compile the source file *gobackn.c*, which has the application and *ASSIST*(assist.c) and link them to provide the executable and run it as follows:

```
#>gcc -c assist.c
#>gcc -c gobackn.c
#>gcc -o gobackn gobackn.o assist.o -lpthread
#>./gobackn
```

After the execution the output file *outfile* is the copy of the input file *infile*.

C.2 Spanning Tree Code

C.2.1 Data Structures and Initialization

```
#define PROBE     0    // message types
#define ACK       1
#define REJECT    2
#define IDLE      0    // states
#define XPLORD    1
#define TERM      2
#define N         6    // number of nodes
#define ROOT      5    // root node

int parents[N]={0}, currstates[N]={0};
int neighs[N][N+3]={0}, others[N][N+3]={0},
                childs[N][N+3]={0};

typedef int  (*fnptr_t)();
typedef fnptr_t fsm_t[2][2];
fifo_t fifos[N];
pool_t mypool;

void init_sys() {int i;
                for(i=0;i<N;i++)
                { init_fifo(&(fifos[i]));
                  childs[i][0]=-1;
                  others[i][0]=-1;}
                init_pool(&mypool,20);
        }
```

C.2.2 Tree Construction Thread

```
/***********************************************
            CRobs Thread
 ***********************************************/
```

```
void CRobs_node(int *mp)
{       int i,j,me=(int )*mp;
        bufptr bp;
        fsm_t myfsm;

        currstates[me]=IDLE;

        myfsm[0][0]=act00;
        myfsm[1][0]=act10;
        myfsm[1][1]=act11;
        myfsm[1][2]=act12;

// Configure my neighbor list according to my id

switch (me){
case 0 : neighs[me][0]=5;neighs[me][1]=1;neighs[me][2]=-1;
           neighs[me][N+2]=2; break;
case 1:   neighs[me][0]=2;neighs[me][1]=4;neighs[me][2]=0;
           neighs[me][3]=5; neighs[me][4]=-1;neighs[me][N+2]=4;
           break;
case 2 : neighs[me][0]=1; neighs[me][1]=3;neighs[me][2]=4;
           neighs[me][3]=-1; neighs[me][N+2]=3; break;
case 3 : neighs[me][0]=2;neighs[me][1]=4;neighs[me][2]=-1;
           neighs[me][N+2]=2; break;
case 4 : neighs[me][0]=1;neighs[me][1]=3;neighs[me][2]=5;
           neighs[me][3]=2; neighs[me][4]=-1;neighs[me][N+2]=4;
           break;
case 5 : neighs[me][0]=0;neighs[me][1]=4;neighs[me][2]=1;
           neighs[me][3]=-1; neighs[me][N+2]=3;
           break;
}

  if (me == ROOT){                              // if I am root do
        for(i=0;neighs[me][i]!=-1; i++) // send PROBE to N(i)
        {
          bp=get_buf(&mypool);
          bp->sender=me,
          bp->type=PROBE;j=neighs[me][i];
          write_fifo(&fifos[j],bp);
        }
          currstates[me]= XPLORD;
    }

while (currstates[me]!=TERM)
                                        // loop until state = TERM
    {
      bp=read_fifo(&fifos[me]);
      (*myfsm[currstates[me]][bp->type])(me,bp);
    }
```

```
            // output parent and children
            printf(" I am %d my parent is %d \n ", me,
                                            parents[me]);

            for(i=0;childs[me][i]!=-1; i++) // send PROBE to N(i)
               printf("my child is %d  ", childs[me][i]);

}
```

C.2.3 Actions

```
/*******************************************************
                act00 :        PROBE received first time
*******************************************************/

act00(int id, bufptr bp) {
        int i,j;
        parents[id]=bp->sender; // mark sender as parent
        others[id][N+2]=1;
        put_buf(&mypool,bp);    // return buffer to pool
        for(i=0;neighs[id][i]!=-1; i++) // send PROBE to N(i)
        { if (neighs[id][i]==parents[id]) continue;
             bp=get_buf(&mypool);        // skip sender
          bp->sender=id;
          bp->type=PROBE; j=neighs[id][i];
          write_fifo(&fifos[j],bp);}
          currstates[id] = XPLORD;
   }

/*******************************************************
            act11 : reject PROBE as parent exists
*******************************************************/

act10(int id, bufptr bp){
        bp->type=REJECT;
        bp->sender=id;
        write_fifo(&fifos[bp->sender],bp);
        }

/*******************************************************
            act11 :        ACK received
*******************************************************/

act11(int id, bufptr bp ){
        int i=childs[id][N+1];
        childs[id][i]=bp->sender; // include sender in childs
        childs[id][N+1]++;            // increment index
```

```
        childs[id][N+2]++;              // increment count
        if(neighs[id][N+2]==childs[id][N+2]+others[id][N+2]) {
           bp->type=ACK;             // if all neighbors responded
           bp->sender=id;            // send ACK to parent
           write_fifo(&(fifos[parents[id]]),bp);
           currstates[id]=TERM;
           childs[id][i+1]=-1;
              }
        else
           put_buf(&mypool,bp);   // return buffer to pool
     }


/*********************************************************
            act12 :        REJECT received
*********************************************************/

act12(int id, bufptr bp){
        int i=others[id][N+1];
        printf("\n --actme : %d act12  i : %d--", id,i);
        others[id][i]=bp->sender; // include sender in others
        others[id][N+1]++;     // increment index
        others[id][N+2]++;     // increment count
        if(neighs[id][N+2]==childs[id][N+2]+others[id][N+2]){
             bp->type=ACK;            // if all neighbors responded
             bp->sender=id;        // send ACK to parent
             write_fifo(&(fifos[parents[id]]),bp);
                 currstates[id]=TERM;
              }
        else
           put_buf(&mypool,bp);    // return buffer to pool
     }
```

C.2.4 The Main Thread

```
main(){
        pthread_t tids[N]; int i;
        init_sys();
        for(i=0;i<N;i++)
           pthread_create(&tids[i],NULL,(void*)CR_node,&i);
        for(i=0;i<N;i++)
           pthread_join(tids[0],NULL);
}
```

Index

K. Erciyes, *Distributed Graph Algorithms for Computer Networks*,
Computer Communications and Networks, DOI 10.1007/978-1-4471-5173-9,
© Springer-Verlag London 2013

Printed in the United States
By Bookmasters